Chinese Forest Fire Control

Chief editor Zhai Hongbo

China Agricultural Science and Technology Press

图书在版编目（CIP）数据

中国森林防火 = Chinese Forest Fire Control : 英文 / 翟洪波等编著. —北京：中国农业科学技术出版社，2019. 6

ISBN 978-7-5116-4031-4

Ⅰ. ①中… Ⅱ. ①翟… Ⅲ. ①森林防火—中国—英文 Ⅳ. ①S762.3

中国版本图书馆 CIP 数据核字（2018）第 020499 号

责任编辑　金　迪　崔改泵
责任校对　贾海霞

出 版 者　中国农业科学技术出版社
　　　　　北京市中关村南大街12号　　邮编：100081
电　　话　（010）82109194（编辑室）（010)82109702（发行部）
　　　　　（010）82109709（读者服务部）
传　　真　（010）82106636
网　　址　http: // www.castp.cn
经 销 者　全国各地新华书店
印 刷 者　北京建宏印刷有限公司
开　　本　787mm × 1092mm　1/16
印　　张　10
字　　数　254千字
版　　次　2019年6月第1版　　2019年6月第1次印刷
定　　价　55.00元

Chinese Forest Fire Control

Asia-Pacific Network
for Sustainable Forest Management and Rehabilitation

Chief editor: Zhai Hongbo

Editors: Wei Xiaoxia Liu Guihong

Revisers: Benjamin Forrest Sun Rui Hu Chuyu

Preface

The Asia-Pacific Network for Sustainable Forest Management and Rehabilitation (APFNet) is a regional organization dedicated to advancing sustainable forest management and rehabilitation in the Asia-Pacific region. The proposal to establish APFNet was made by the People's Republic of China and co-sponsored by Australia and the United States, and was adopted by Asia-Pacific Economic Cooperation (APEC) Leaders at the 15th APEC Economic Leaders' Meeting in 2007.

APFNet has 31 members, including 26 member economies in the Asia-Pacific region and 5 international organizations. China is one of the 26 member economies.

China stretches across a vast area boasting complex natural conditions, rich bio-species and a great diversity of community types. The economy now possesses approximately 208 million hectares of forests and 53 million hectares of wetlands. Nevertheless, it is an economy with inadequate forest resources, vulnerable ecology and scarce ecological products. The forest coverage for the whole economy currently stands at 21.63%, only equivalent to 70% of the world average level, whilst the per capita forest area is only 1/4 of the world average, and the forest growing stock per capita is only 1/7 of the world average.

As one of the principal parts of the terrestrial ecosystem, the forest represents the largest terrestrial biome on earth. It is considered to be the cradle of humankind and also an essential natural resource for humankind. Forest fire, ranking among the world's eight major natural disasters, seriously endangers the safety of people's lives and property, and forest resources and even triggers ecological disasters.

Since the founding of the People's Republic of China, the Chinese government has been highly valuing forest fire control, which is the premise and foundation for protecting forest resources by law, improving the ecological environment and boosting the forestry's sustainable development. As a result of the increasing investments in forest fire control, China has made considerable progress in this aspect. It has been an irresistible trend for fire control to be managed in a legal, scientific and professional manner.

In this book, we introduced Chinese experience and approaches for forest fire control. We hope it is helpful and instructive to other economies.

APFNet Executive Director

Contents

CHAPTER I

Introduction

I. Basic Concepts of Forest Fire Control

1. Forest Fire Control

The term "Forest Fire Control" referred to in this book is the synonym of "Forest Fire Prevention" as we always mentioned. It is the practice of guarding against forest fires, preventing forest fires from spreading, suppressing or extinguishing forest fires along with fire investigation, loss evaluation and archives management.

2. Forest Fire

Forest fire is a free-burning fire that occurs on forest land, generally including controlled and uncontrolled fires. A controlled fire is typically a planned burning of forest combustibles in a specific area of a selected forest land; whilst an uncontrolled fire may develop into a harmful forest fire.

3. (Harmful) Forest Fire

It is an uncontrolled fire on forest land that is free-burning, exists beyond a specific area and causes some losses.

4. Classification of Forest Fires

As stated in the *Regulations on Forest Fire Control* promulgated by the State Council on January 16th, 1988 and revised and adopted at the 36th Executive Meeting of the State

Council on November 19th, 2008, forest fires shall be classified by affected forest area and casualty into the following four types:

(1) Ordinary forest fire: it refers to a fire that has affected less than 1 ha area or burned in other forests, or causes at least 1 but less than 3 deaths or at least 1 but less than 10 serious injuries.

(2) Serious forest fire: it refers to a fire that has affected at least 1 ha but less than 100 ha area, or causes at least 3 but less than 10 deaths or at least 10 but less than 50 serious injuries.

(3) Fatal forest fire: it refers to a fire that has affected at least 100 ha but less than 1,000 ha area, or causes at least 10 but less than 30 deaths or at least 50 but less than 100 serious injuries.

(4) Devastating forest fire: it refers to a fire that has affected more than 1,000 ha area, or cause at least 30 deaths or at least 100 serious injuries.

The term “at least” mentioned in the preceding paragraphs shall include the number itself, and the term “less than” shall not.

5. Forest Combustion

It is a natural phenomenon where forest combustibles oxidize intensely, release heat and sparkle at certain temperatures, and also a complex physico-chemical process.

6. Three Elements of Forest Combustion

Combustible, comburent (oxygen) and desired temperature constitute the three elements of combustion. The same is true with forest combustion. However, forest combustion often proceeds in an open system, making it unaffected by oxygen condition. Therefore, meteorological condition has an enormous impact on the combustion process.

As far as fire control is concerned, forest combustible, ignition source and meteorological condition (fire weather) are known as the three elements of forest combustion.

7. Forest Combustible

Forest combustibles refer to all organic substances in the forest that can combust, including trees, bushes, vines, vegetation, mosses, lichens, surface litters (overturned trees, coarse woody debris, branches, fallen leaves, flowers and fruit) and underground humics and peats.

8. Forest Combustible Category

It refers to a complex of combustibles that occupy some spaces in the forest, stay relatively stable within a certain period of time and have identical or similar combustibility. In general, combustibles of the same category are almost identical with the combustible type, load, arrangement and compactness.

9. Ignition Source

It refers to a heat source that induces forest combustion, which is divided into natural and artificial ignition sources. Natural ignition sources include lightning, volcanic eruption, meteorite falling, rolling stock sparkling, peat spontaneous combustion, etc.; while artificial ignition sources include fires for productive and living activities, cross-border fires and intentionally-set fires.

10. Fire Environment

Fire environment is a totality of all factors underlying the outbreak and spreading of forest fires except combustible and ignition sources, mainly including weather, climatic, terrain and soil conditions and forest microclimate.

11. Meteorological Conditions (Fire Weather)

Meteorological conditions refer to the general characteristics of all meteorological elements at a given time or within a given time frame in a given place, i.e. such atmospheric conditions as coldness and warmth, cloudiness and brightness, dryness and wetness, snow and rain, wind and cloud and their changes with time in a specific area. The meteorological factors closely related to forest fires mainly include air temperature and humidity, rain and wind.

12. Forest Combustion Process

Forest Combustion Process is a complex physico-chemical process that comprises preheating, thermal decomposition, flaming and extinguishing stages.

13. Forest Fire Behaviors

Forest fire behaviors depict various phenomena and features found with a forest fire in the whole process where forest combustibles are ignited, extend and spread and eventually die out. These behaviors include rate of fire-spread velocity, burning intensity, flame height, fire severity, convection column, fire whirl, heat wave, etc. Meanwhile, fire-spread velocity, burning intensity and flame height are considered the three major measures of forest fire behaviors.

14. Forest Fire Spread

After forest combustibles are ignited by an ignition source, forest fire may spread around; such combustion phenomenon is known as the forest fire spread. It involves the velocity at which the fire spreads around and the shape changes, expansion rate and circumference growth rate of the fireground.

15. Fireground

Firegrounds of varied sizes and shapes will be produced depending on the states of forest combustibles, fire environment (mainly referring to topography and wind), fire duration and other factors. A fireground is divided into fire head, fire flank, fire rear and fire slash based on the directions of fire spread and burning behaviors at different positions of the fireground.

16. Fire Intensity

Forest intensity represents the rate of heat released by forest combustibles. Fireline intensity, reaction intensity and fire front intensity are the three common measures of fire intensity. In forest fire control practices, fireline intensity is often used as decision reference for setting or suppressing a fire. It is measured as the heat or energy released across per unit length of fireline per unit time. Fireline intensity generally ranges from 20 to 10,000 kJ/m. Forest fires with fireline intensity of <750 kJ/m, 750–3,500 kJ/m and >3,500 kJ/m are respectively considered low intensity, moderate intensity and high intensity fires. Low and moderate intensity fires can be put out directly by firefighters, while high intensity fires must be suppressed indirectly.

17. Fire Severity

Fire severity is dependent upon the product of the rate of energy released per unit area and fire residence time. In practice, it is often measured as percentage of total trees burned by fire.

18. High-risk Forest Fire Behaviors

High-risk forest fire behaviors include spotting, fire whirl, firestorm and heat wave.

(1) Spot fire: a fire started by burning combustibles that are carried high up by intense convection columns formed by high-energy main fire and flying downwind from the fire-ground.

(2) Fire whirl: a whirlwind occurring when intense convection and lateral wind conditions contribute to form whirling eddies of air in the combustion area.

(3) Firestorm: a natural phenomenon occurring when plenty of spot fires are produced and an explosive combustion is created by burning and aggregation of many sparks, thereby transforming low flames into a sheet of flames.

(4) Heat wave: an invisible high-temperature and high-velocity gas flow (intense heat advection) formed when a great many of combustibles are burning intensely and releasing

tremendous heat to heat up surface air.

19. Forest Fire Combustion Types

Forest fires fall into four types including ground fire, surface fire, crown fire and stand fire, depending on the flaming position, rate of spread, fire intensity, flame height and propagation continuity.

(1) Ground fire: a fire fed by underground combustibles such as humics, peats and roots. It typically emits a small amount of smoke but no flame on the ground. And it is characterized by slow rate of spread, high temperature, heavy disruption and long duration.

(2) Surface fire: a fire fueled by surface fuels and spreading along wildland surface. It is the most common forest fire behavior, characterized by a dual nature. A serious surface fire may damage seedlings, underwood, bushes and exposed roots, induce plant diseases and insect pests and even destroy vast forests. However, ordinary surface fire would help remove forest combustibles effectively and optimize soil physicochemical properties.

(3) Crown fire: a fire burning at the canopy level, characterized by fast rate of spread, high temperature, heavy disruption and great suppression difficulty. A swiftly advancing crown fire may move with the wind at a rate of 8–25 km/h or above, while a steadily advancing crown fire may spread with the wind at a rate of 5–8 km/h.

(4) Stand fire: a fire that heats and ignites the trunks of trees.

20. Forest Fire Forecast

Forest fire forecast is the practice of measuring, calculating and determining the likelihood of forest fire and prospective fire behaviors through collecting, observing, and recording all the data and materials concerning the occurrence and development of forest fire.

21. Forest Fire Risk Rating

As stipulated in the *Rank of the Regionalization on Nationwide Forest Fire Risk* (LY/T 1063–2008), all county-level administrative units across China are grouped into the following four forest fire risk ratings, depending on such factors as resource conditions, weather conditions, tree species (groups) combustion type, population density and road network density: Level 1–Extreme risk; Level 2–high risk; Level 3–moderate risk; and others–low or no risk.

22. Forest Fire Monitoring Hierarchies

Forest fire monitoring is conducted at four hierarchies, including ground patrol, watch-tower fixed observation, UVA (unmanned aerial vehicle) patrol and satellite monitoring.

23. Forest Guard

As stipulated in the *Forest Law of the People's Republic of China*, forest guards may be appointed by people's governments at the county or township level. The main duties of forest guards shall be: patrol the forests and prevent forest resources from being destroyed. Their main task is ground patrol.

24. Ground Patrol

The act of trained forest guards and policemen patrolling on the ground along certain routes by walking or by means of transportation (horse, bike, motorbike, powerboat, etc.).

25. Watchtower Observation

Watchtower observation is a forest fire monitoring method in which the fire source is located and reported timely by arranging a watchtower high above the ground or using the top floor or upper structure of an existing high-rise building as the observatory platform for fixed-point monitoring of forest fires. The watchtower position is generally selected in the following 3 ways: topographic modeling, map locating and field reconnaissance. A watchtower must be well equipped with basic facilities and equipment such as living installations, observation instruments, locator devices and communication equipment.

26. Watchtower Monitoring Coverage

The percentage of woodland area covered by watchtower to the total woodland area (excluding the overlapping area between watchtowers or video monitoring systems).

27. Aerial Patrol

Aerial patrol is an important means of forest fire control, whereby UAVs are flying over the forest along the pre-set course for patrolling purpose. By patrolling, UAVs monitor forest fires, locate fireground or fire source and report relevant information to the aircraft base or forest fire control command center.

28. Satellite Remote Sensing Monitoring

In satellite remote sensing monitoring of forest fires, the spectral or microwave sensor arranged on the artificial satellite space platform is used to scan or receive the spectral or microwave information sent from ground features on the earth and transmit the same to the ground receiving station; the ground receiving station then detects the heat or fire source after enhancement processing of the received information by the Image Data Processing System (IDPS), followed by tracking monitoring. Satellite remote sensing monitoring of forest fires has the advantages of large monitoring coverage, high frequency, all weather operation, high speed and accuracy, making it possible for real-time monitoring of forest fires.

29. Forest Fire Weather Danger Rating

Forest fire weather danger is divided in to five levels as follows.
(1) Level 1: Negligible; forest combustibles cannot be ignited.
(2) Level 2: Low; forest combustibles are unlikely to be ignited.
(3) Level 3: Moderate; forest combustibles are fairly susceptible to ignition.
(4) Level 4: High; forest combustibles are susceptible to ignition.
(5) Level 5: Extreme; forest combustibles are highly susceptible to ignition.

30. Raw Soil Belt

A belt of raw soil exposed on the surface following mechanical or manual removal of vegetation. It is mainly used to prevent low-intensity surface fires from spreading. And it is typically constructed before approaching of fire season.

31. Firebreak

A barrier to stop the progress of bushfires constructed along the pre-set route by removing trees, shrubs and weeds by burning, using chemical agents mechanically or manually. Firebreaks can be divided into frontier firebreak, railway firebreak, boarder firebreak and forest firebreak based on their locations; firebreaks can be classified into primary firebreak and secondary firebreak depending on their width and function.

32. Forest Fire Breaking Network

Various forest fire barriers linked and acting together in a network, comprising artificial and natural barrier systems. Artificial barriers mainly include road, raw soil bet, fire trench, firebreak, firebreak tree belt, firewall, forest edge or removed undergrowth vegetation belt, farmland, etc.; natural barriers mainly include river, lake, pool, rocky area, sand dune and all natural topographic conditions that contribute to stopping the spread of forest fires. Natural barriers are independent of artificial ones and lack of continuity. But in order to make natural barriers' more effective in stopping the spread of forest fires, it is necessary for the two to act together to form a forest fire breaking network.

33. Woodland Road Network Density

The ratio of the area of rated roads in the woodland to the total woodland area, which is typically measured in m/ha.

34. Aerial Ranger Station

Aerial ranger station is an important part of aerial forest firefighting and a specialized unit engaging in rescue and relief work. Its main task is to prevent and suppress forest fires. And its main functions include: undertake aerial forest firefighting in its patrolling area; perform aerial forest firefighting service contracts; organize, arrange, coordinate and direct all aerial forest firefighting activities carried out by the station, render logistics service and provide aircraft ground support for aerial ranger stations having ground supporting functions in accordance with relevant national laws and regulations; collect and transmit data, report flight and fireground dynamics; draw up aerial firefighting schemes and put forward suggestions on forest fire suppression; complete flight route planning and technical innovation, launch scientific research programs on aerial forest firefighting, and develop and facilitate wider application of novel aerial forest firefighting techniques; and finish other tasks assigned by superior sectors.

35. Mobile Aerial Ranger Station

Mobile aerial ranger station is opposite to fixed ranger airport or station. In a fixed aerial ranger station, flight missions for aerial forest fire control purpose are accomplished with the aid of ground support. While for a mobile aerial ranger station, the concept of the station's ground support functions is extended and developed to integrate all ground support facilities and equipment relating to command, communication & navigation, fuel and weather in a scientific way. The mobile aerial ranger station can be moved as close as possible to the helicopter's operating area at any time as needed so as to protect flight safety anytime anywhere, minimize ineffective flights, increase operation time and opportunities per unit time and cut down the costs and expenses of aerial forest fire control.

36. Fleet Spray Firefighting

Fleet spray firefighting is the use of a fleet of fixed-wing aircraft carrying chemical agents to spray liquid directly toward fire head and fireline or spray liquid toward firebreak at forepart of fire head and fireline for the purpose of fire suppression and extinguishment.

37. Hanging Bucket Firefighting

Firefighting helicopter dropping a water spray from its specialized hanging buckets carrying water or chemical agents onto the fire. Firefighting with helicopter buckets contains direct and indirect fire suppression. In a direct fire suppression, helicopter buckets are filled with water to spray water directly onto the fire; while in an indirect fire suppression, helicopter buckets skimming water are dropped from the helicopter to the ground flexible pond for use by the ground firefighting crew to fire suppression and mopping up.

38. Rappel (fast-rope) Firefighting

It is an advanced and effective direct fire suppression method applied in aerial forest firefighting. Winch and steel rope are used in rappel firefighting, while rope controller and rope are used in fast-rope firefighting. Specifically, firefighting crew on board the flying helicopter is ferried and dropped safely and rapidly using harness systems and other equipment to the ground in vicinity of the fire to perform firefighting duties.

39. Helitack

Helitack is the use of helicopters that can take off and land in the field to deliver firefighting crew, tools and equipment timely to the ground in vicinity of the fire for encirclement, organize and direct firefighting activities and perform firefighting force adjustment and reassignment uninterruptedly.

40. "Four Firsts + Two Guarantees" Principle of Firefighting

When directing the acts of attacking a fire, commanders of all levels must follow the "Four Firsts + Two Guarantees" principle of firefighting to put out the fire swiftly and effectively. Four Firsts: suppress fire head first, suppress grass pond fire first, suppress flaming fire first and suppress perimeter fire first; Two Guarantees: guarantee link-up operation, and guarantee no-reignition.

41. Shortwave Communication

Shortwave communication is a radio transmission through ground and sky waves using shortwave frequencies of 3–30 MHz (10–100 m wavelength). Ground waves can propagate dozens of kilometers, and sky waves can travel up to 10,000 km.

Features:
(1) Multiple reflections by ionosphere or between ionosphere and ground allow electromagnetic waves to pass through giant barriers and propagate far away.
(2) Simple and low-cost communication equipment.
(3) Subject to the influences of diurnal and temporal variations and solar storm.
(4) The skip distance found with; onosphere reflection affects the communication reliability and success rate.
(5) Jamming radio stations lead to serious interference and vague voice.

42. USW (ultrashort wave) Communication

Intercom is a typical USW communication device used in communication of forest fire control. It has the advantages of portability, small power consumption, low cost and good communication quality, thus it is widely applied in forest fire control. However, the line-of-sight propagation of USW communication is highly affected by topography and ground features, resulting in a short propagation distance. So, it is more suitable for plains or gently undulating places.

43. Microwave Communication

Microwaves are electromagnetic waves like radio waves, infrared ray and visible light. Microwave waves are electromagnetic waves with frequencies ranging from 300 MHz to 300 kMHz. Microwave waves have higher frequencies than general radio waves and are thus known as "UHF electromagnetic waves". Wave-particle dualism, which is typical of electromagnetic waves, also applies to microwaves, exhibiting three features including transmittance, reflection and absorption.

Microwaves travel in a line-of-sight fashion like ultrashort waves for a short distance of 50–60 km. Microwave relay stations must be erected to transmit signals further. For this reason, microwave communication is also known as microwave relay communication. It is supplementary to satellite communication and shortwave communication, characterized by large coverage, wide band and high transmission quantity. Moreover, it can be networked with computers.

44. Satellite Communication (incl. maritime satellite telephone)

Satellite communication is a kind of microwave communication using satellite as relay station. Satellites hanging high above the earth are seldom affected by topography and ground features. This mode of communication has the merits of long communication distance, large coverage, high communication capacity and effectiveness. So it is widely used throughout the world. As to the "3S" technology developed for forest fire control, data about forest vegetation, heat source or forest fire is captured by satellites and transmitted to ground receiving station through microwave communication so as to locate fire source or fireground and realize real-time monitoring of fireground.

45. Relay Link Processor

Relay link process is a critical communication device used for networking of multiple relay stations operating in the same network or interconnection and interworking of multiple regional relay communication subsystems in the same network. In practice, if the local communication system uses 150 M band, then the link processer should use 400 M band; if the local communication system uses 400 M band, then the link processer should use 150 M band. In exceptional cases, different relay communication systems in the same network may use link processors of 150 M and 400 M bands simultaneously for interconnection and interworking.

46. Fireground Real-time Multimedia Information Aircraft-based Transmission System

In this system, aircrafts for forest fire control act as the communication platform to realize fireground image collection, fireground situation plotting and voice communication with ground commanders in a real-time fashion and to transmit such images and plots to the ground command in a real-time manner with the aid of aircraft-based USW transmitter-receiver. In this way, visual and reliable integrated information is made available to the fire frontline command post as the basis for command decision.

The system uses aircrafts about forest fire control to provide fireground images, situation maps and other relevant information about fire frontline command posts located in a place with complex geographic and geomorphic conditions. It is particularly suitable for fire frontline command posts stationed in gorge areas in southern China, where helicopters can't land to support communication of command decision and information transmission.

47. Communication Coverage

Communication coverage is the percentage of the area covered by all wireless (excl. shortwave communication) and wire communication media available to the total area of the target region.

48. "Four Networks + Two Musts"

"Four Networks + Two Musts" is an important part of facilities construction for forest fire control. "Four Networks" include forest fire prediction and forecast network, communication network, watchtower observation network and firebreak network; and "Two Musts" denote firefighting crew must be trained specialists and firefighting tools must be mechanical devices.

49. Forest Fire Investigation

As an important part of modern forest fire management, forest fire investigation is a key link of post-disaster management after a forest fire is extinguished. It is an investigation of the cause of the forest fire, the person who inflicts the incident, total area of affected forest and forest stock, casualties and other economic losses organized by forestry administrations of people's governments at county level and above together with competent authorities.

50. Forest Fire Archives

Forest fire archives are the original records of fire outbreak, spreading, suppression and extinguishment, fire investigation and analysis as well as fire case settlement. The archives can be used as:

(1) The basis for assessing the performance of the forest fire department.

(2) The basis for determining the fire nature and ascertaining the responsible person.

(3) The basis for summarizing the regularities of forest fire outbreak and enhancing prediction and early warning performance.

(4) Publicity and educational materials for forest fire control.

II. Forest Fire Characteristics

1. Attributes and Dual Nature of Forest Fire

Forest fire is a highly destructive burst natural disaster subject to many human factors, with great response and relief difficulties. It occurs independent of man's will and is unlikely to be completely eradicated. Widespread forest fire is listed as one of the eight major natural disasters in the world by the United Nations Food and Agriculture Organization (FAO).

Meteorological condition is the first and foremost condition for forest burning, followed by vegetation (different vegetations vary in combustibility) and source of ignition. The source of ignition in flammable forest stands can start a forest fire only in drought, windy, high-temperature and low-humidity weather conditions. Therefore, forest fire is classified as meteorological fire to a large extent.

Worldwide forest fire control practices prove us that the longer a forest stays unaffected by fire, the more forest combustibles will accumulate and the more likely a fatal or devastating forest fire will occur. When forest combustibles accumulate to some extent, they must be burnt away; otherwise, they will present a hazard to the forest.

Forests need fire since it is the fuel for forest succession. Fire disrupts forest resources, fauna and flora and causes water and soil erosion and air pollution, but it helps to maintain biodiversity and landscape diversity, improve the survival conditions for some animals and plants, enhance soil fertility and hasten plant growth, blooming, fruiting, gum yield and regeneration. Human beings should prevent the harms of forest fire while taking advantage of the benefits of forest fire.

2. Characteristics of Forest Fires in China

(1) High frequency and heavy losses. In 1950–2005, about 741,897 forest fires took place, averaging 13,489 cases per year, ranking the 5th in the world; the fire-sweeping areas totaled 3,940 ha, averaging 716,400 ha per year. 29,759 injuries and 5,245 deaths were caused by forest fires, averaging 541 injuries and 95 deaths per year. In 2009–2013, China suffered on average 6,525 forest fires every year, resulting in 33,000 ha fire-sweeping areas and 83 casualties. These figures declined by 34%, 79% and 38% respectively compared with those of 2001–2008. The rate of affected forest was controlled at 0.1% or below, thus the forest resources were effectively protected and the social stability of woodlands was well maintained.

(2) Wide and concentrated distribution. Every province (autonomous region and mu-

nicipality) suffers forest fires, but the areas that are most targeted by forest fires are Northeast China, Southwest China, Central China and Southeast China. In 2009–2013, the top 10 provinces with most cases of forest fire were Guizhou, Hunan, Henan, Hubei, Guangxi, Sichuan, Yunnan, Fujian, Zhejiang, Guangdong; statistics of forest fire distribution in different forest fire control zones show that Southwest China comes on the top list, taking up 41.4% of the total of the nation, followed by Central China (41.4%), Sountheastern China (35.86%), Northeastern China (13.37%), Northwestern China (4.08%) and Northern China (3.39%). Among the provinces (autonomous regions) that are mostly attacked by forest fires, forest fires concentrate on a few regions (municipalities, prefectures and leagues) or counties (municipalities, districts and banners). For example, forest fires concentrate on Daxinganling and Heihe regions in Heilongjiang Province; Hulunbeier City and Hinggan League in Inner-Mongolia Autonomous Region; Simao, Lincang, Lijiang, Honghe and Diqing in Yunnan; Baise, Hechi and Nanning in Guangxi Zhuang Autonomous Region; Hengyang and Shaoyang in Hunan; and Longyan and Ningde in Fujian.

(3) Artificial source of ignition is the dominant source of ignition. In the forest fires with identified cause and origin, the fires started by artificial source of ignition make up 90% of the total fires. Lightning is the major natural source of ignition, and lightning fires are mainly distributed in the woodlands in Great Khingan and northern Lesser Khingan Mountains as well as the woodlands in Xinjiang Tianshan and Altai Mountains. Lightning fires and wildfires are rarely found in the forests in South China.

(4) Forest fires are not subject to obvious seasonal variation. In Northeast China forests, fires normally happen in fall and spring; in Xinjiang Uygur Autonomous Region, forest fires mostly happen in summer; in South China forests, fires often take place in winter and spring. The frequency of forest fires increases from south to north in spring and from north to south in fall.

(5) The most frequently fire-hit time exhibits periodicity. It goes in a circle within 5–6 years or of about 10 years. In 1951–2013, there were eight forest fire high-incidence periods across the nation, with each period lasting for 2–3 years.

(6) Forest fire has been on a downtrend. By comparing the incidence of forest fires be-

fore (1950–1987) and after (1988–2000) the outbreak of the record-breaking 1987 fire in Greater Khingan Range, we can find that the average fire cases per year were only 42% of those before the fire, the fire-affected forest area in an average year was only 9% of that before the fire, the fire-sweeping area per incidence was only 21% of that before the fire, the fire incidences per 100,000 ha of forests was only 42% of that before the fire, and the rate of affected forest was only 8.9% of that before the fire. In 2003–2011, the forest fire incidences, fire-sweeping area and resultant casualties declined by 26%, 80% and 9% on average compared with those in 1950–2002. The incidences of ordinary fires and serious fires in 2013 dropped by 0.93% and 100% respectively compared with those in 2012, the fire-sweeping area decreased by 1.61%, and the firefighting cost reduced by 73.29%.

Ⅲ. Development History of Forest Fire Control in China

The modern forest fire control mechanism can be dated back to 1949 when People's Republic of China was founded. Since the founding of new China, the economy's forest fire control mechanism underwent the following 5 phases.

1. Inception Phase (1949–1956)

In October 1950, the People's Government of Northeast China organized an armed ranger force as a prelude to establish the workforce for forest fire control. In 1951–1956, the CPC Central Committee and the State Council (Government Administration Council) published a package of normative documents such as the *Instructions to Party Committees of All Levels on the Issues Concerning Forest Fire Prevention*, the *Instructions on Forest Fire Prevention* and the *Emergency Notice on Forest Protection and Fire Control*, requiring all regions involved, especially the key state-owned forest areas, to set up a forest fire department. In 1952, the former Ministry of Forestry begun to set up forest fire control bases in the woodlands in Northeast China and Inner Mongolia. By 1952, 18 provinces (autonomous regions) had established forest fire control command organs. Statistics show that the total fire-sweeping forest area decreased by 74.1% in 1952 compared with 1951. In August 1956, the former Ministry of Forestry convened the Workshop on Forest Fire Control Science and Technology in the Woodlands in Northeastern China and Inner Mongolia. It was stressed on the symposium that gross-root forest management organs should be established in the woodlands to implement forest fire control technologies and measures in a well-organized manner while strengthening the public's involvement in forest fire control.

2. Full-scale Development Phase (1957–1965)

In January 1957, the former Ministry of Forestry organized the Forest Fire Control Office responsible for administration of forest fire control activities across the economy. Since then, forest fire control stepped into a period of "mass prevention dominated, mass and ranger forest protection combined". Zero forest fire competitions were then in full swing in all counties, districts and villages throughout the economy.

On January 29th, 1960, the Chinese Government entered into the *Agreement on Cooperation in Forest Fire Control* with the Soviet Union Government.

In 1961–1963, the State Council, together with other authorities concerned, promulgated the *Forest Protection Regulations* and the *Trial Procedures for Safe Ignition in Shifting Cultivation, Incineration & Manure Collection and Forestry Production.* These two regulations set out the specific rules on the administration of fires used in the fields in productive activities and became the standards governing nationwide forest fire control work.

The former Ministry of Forestry convened the Working Conference on Forest Fire Control in the Woodlands in Northeastern China and Inner Mongolia in Haila'er City in Inner Mongolia in December 1964, deliberating on the scope and mode of forest fire control in key woodlands.

3. Standstill Phase (1966–1976)

During the period of Great Cultural Renovation, China's forest fire control industry was brought to a standstill. Quite a few of forest fire control organs slipped into a state of paralysis, full-time staff was downsized, construction of forest fireproofing facilities just started in key woodlands was suspended; some facilities were in service for long years out of repair; effective forest fire control regulations and systems came under siege; and excessive deforestation, forest destruction and forest fire were found everywhere. On May 17th, 1966, a devastating fire caused by burning of anthills swept the Songling District in the Greater Khingan. The fire lasted for 46 days and burnt 546,000 ha forest areas. On

September 23rd, 1967, the CPC Central Committee, State Council and Central Military Commission worked together to issue the *Notice on Enhancing the Administration of Forest Protection and Curbing Forest and Tree Destruction*. In 1975 when Deng Xiaoping took charge of the work of the Central Committee, considering the serious forest fires then happened, he convened a field meeting on nationwide forest fire control in Harbin in July to deliberate on the specific measures for preventing forest fires. However, affected by the "Gang of Four", the proposal adopted at the meeting was not put into practice, thereby exacerbating the forest fires. In September 1976, the Forestry Administration of Suiling County in Heilongjiang Province suffered a devastating fire, resulting in 350,000 ha fire-affected areas. 10,737 persons were assigned to attack the fire. The fire lasted for 40 days.

4. Recovery Phase (1977–1986)

Since the 3rd Plenary Session of the 11th Central Committee of the CPC was convened, China's forest fire control industry started to restore vitality. On February 23rd, 1979, the *Forest Law of the People's Republic of China (Trial)* was adopted at the Sixth Meeting of the Standing Committee of the Fifth National People's Congress; on February 9th, 1981, the State Council issued the *Notice on Strengthening Forest Fire Control*; and on March 8th, 1981, the CPC Central Committee and State Council jointly published the *Decisions on Issues Concerning Forest Protection and Forestry Development*. Thereafter, the former Ministry of Forestry held several national and regional working conferences to deliberate on the arrangement of forest fire control work, which brought the establishment and construction of forest fire control organs, specialized staff and facilities to a higher level.

The frequency of forest fires in the woodlands in Northeast China and Inner Mongolia has been on the decrease since 1980. In 1983, the former Ministry of Forestry set up the Editorial Board of *Forest Fire Prevention* magazine and started its publication. On May 4th, 1984, China and Canada entered into the *Agreement on Sino-Canada Cooperation in Forest Fire Control* in Beijing, whereby the two sides agreed to cooperate in establishment of a forest fire control and suppression system in the Greater Khingan Range. It provides reference for the design of forest fire prevention management technologies and systems in other regions in China. In 1986, the Fire Control Specialized Committee of China Fire Control Association was established.

In late March 1986, two forest fires hit Anning County and Yuxi City in Yunnan Province in succession, leading to 80 deaths of firefighting staff, civilians and PLA soldiers and about 100 injuries. In March 24th, the State Council issued the *Emergency Notice on Strengthening Forest Fire Control*, requiring governments of all levels to put fortified efforts in forest fire control and take effective actions to prevent forest fires. On March 31st, the State Council released the *Report on the Serious Forest Fire Situations in Yunnan Province*; later, Yunnan Provincial People's Government submitted an inspection report to the State Council and take specific measures to prevent forest fires. In December 1986, the former Ministry of Forestry held a national working conference on forest fire prevention in Jiujiang City of Jiangxi Province, summing up experience and lessons from serious forest fires happened in spring of the year, exploring the fundamental measures against forest fires and commending and awarding the outstanding units and individual in forest fire prevention in all regions.

5. Historical Transition Phase (1987 till now)

A forest fire took place in the northern woodlands in the Greater Khingan Range in Heilongjiang Province on May 6th–June 2nd, 1987. It is the most devastating forest fire ever reported since the founding of new China, leaving 1,330,000 ha areas and 37,811,000 m^3 standing forest stocks burnt. The fire swept 3 forestry bureaus in one night (on May 7th) and burnt most buildings and facilities in 7 forestry stations. The fire caused 213 deaths, 226 injuries and 10,807 affected households. About 58,800 firefighters, 96 aircrafts, 1,600 vehicles, 3,600 pneumatic extinguishers and 80 trains were assigned to the incident, and the direct economic losses amounted to RMB 500 million.

This devastating forest fire attracted a great deal of attention from the Party and the public. In August 1987, the State Council organized the Central General Headquarters of Forest Fire Prevention which was renamed the National Headquarters of Forest Fire Prevention in 1989, for which the vice-premier in charge acted as the commander-in-chief. On January 16th, 1988, the State Council enacted the *Regulations on Forest Fire Control*; and on November 8th, 1989, the National Headquarters of Forest Fire Prevention, the Ministry of Human Resources and Social Security of the P.R.C. and the former Ministry of Forestry jointly printed and distributed the *Notice on Strengthening the Construction of Forest Fire Control Systems*.

The CPC Central Committee and the State Council have all along attached great importance to forest fire prevention. On April 14th, 2004, the General Office of the State Council released the *Notice on Further Strengthening Forest Fire Control*; on May 14th, 2005, the *National Contingency Plan for Response to Fatal and Devastating Forest Fires* was officially launched by the State Council and took into effect; on November 19th, 2008, the revised *Regulations on Forest Fire Control* was adopted at the 36th Executive Meeting of the State Council; in 2009, the State Council approved the *National Medium and Long-Term Plan for Forest Fire Control* (2009–2015), whereby the economy improved the construction of forest fire equipment and infrastructure in key woodlands and enhanced its ability in forest fire comprehensive prevention and control by increasing the inputs in fire prevention and accelerating the construction of the forest fire prevention, suppression and protection systems. In this way, the rate of affected forest across the economy declined steadily. On December 17th, 2012, the General Office of the State Council issued the *National Contingency Plan for Response to Forest Fires* revised out of the original *National Contingency Plan for Response to Fatal and Devastating Forest Fires*. With this, China has progressed much in forest fire prevention.

In this historical phase, China's forest fire prevention industry experienced 10 major transitions as summarized below: ① from the single experience-based fire prevention to the combined experience and sci-technology-based prevention; ② from the single ignition source control to the combined control of ignition source and fuels; ③ from the single engineered fire prevention to the combined engineered and biological fire prevention; ④ from the single-channel funding to the combined multi-channel fund raising and preferentials application; ⑤ from the single administrative measures-based fire prevention to the combined administrative measures and legal institutions-based fire prevention; ⑥ from the single ground monitoring of forest fires to the combined ground, aerial and space remote sensing monitoring of forest fires; ⑦ from the firefighting based on leaders' instructions to the firefighting based on accountability and firefighting commander systems combined with scientific firefighting methods; ⑧ from the firefighting by the mass to the firefighting by specialized firefighting crews; ⑨ from the single ground firefighting to the combined ground and aerial firefighting; ⑩ from relying solely on domestic forest fire prevention experience to exerting strengths combined with drawing advanced experience from abroad.

The above ten transitions signified that China's forest fire control industry has ushered in a new development stage.

CHAPTER II

Forest Fire Comprehensive Prevention

I. Administration of Forest Fire Prevention

In China, forest fire prevention activities are managed under the responsibility system of administrative chiefs of local people's governments at different levels. Governments and departments at different levels must establish and maintain forest fire prevention organizations to proactively implement the guideline of "prevention first, effective suppression followed" governing forest fire prevention activities.

As prescribed in the *Regulations on Forest Fire Control*, local people's governments at county level and above shall set up forest fire prevention command centers as needed, responsible for organizing, coordinating and directing all forest fire prevention activities carried out in the administrative region. Forestry administrations of local people's governments at county level and above shall take in charge of supervision over and management of forest fire prevention in the administrative region and assume the day-to-day duties of forest fire prevention command centers. Other authorities of local people's governments at county level and above shall fulfill their duties of forest fire prevention based on the assignment of responsibilities.

The organization system of forest fire prevention in China includes the National Headquarters of Forest Fire Prevention, forest fire prevention commands of local people's governments at different levels, gross-roots forest fire prevention organizations, regional forest fire prevention cooperative organizations and specialized firefighting agencies.

Ⅱ. Publicity and Education on Forest Fire Prevention

Over 90% of forest fires in China are started by artificial sources of ignition, most of which are caused by negligence in use of fires. Therefore, it is of great importance to carry out nationwide publicity and education on forest fire prevention and enhance the public's fire prevention awareness so as to effectively prevent forest fires.

1. Major Forms of Publicity and Education

(1) Release authoritative documents, notifications and orders.
(2) Launch fire prevention awareness month or week events during fire season.
(3) Hold all sorts of meetings and assemblies.
(4) Organize forest fire prevention knowledge contests and essay contests with awards.
(5) Compile and print various publicity materials.
(6) Put up permanent propaganda signs.
(7) Make use of modern media for propaganda.

2. Contents of Publicity and Education

(1) The importance of forest fire prevention, and the hazards and risks of forest fires.
(2) The guidelines, policies, laws, regulations and local rules and regulations regarding forest fire prevention enacted by the Party and the State.
(3) Outstanding individuals and units, exemplary deeds and advanced experience found in forest fire prevention.
(4) Analysis of typical forest fire prevention offence cases.
(5) Forest fire prevention knowledge.

Ⅲ. Comprehensive Preventive Systems

1. Responsibility System of Administrative Chiefs

According to the *Regulations on Forest Fire Control* and relevant documents, forest fire prevention activities in China are managed under the responsibility system of administrative chiefs of people's governments at different levels, whereby governors, mayors, county magistrates and village (township) heads are held accountable for all forest fire prevention activities in their administrative regions. Forestry administrations at different levels are responsible for the supervision over and management of forest fire prevention, and all units in the woodland should implement the responsibility system of department and sector leaders under the local government's leadership.

2. Access Control System

As stated in the *Regulations on Forest Fire Control*, the management organs of state-owned forest areas approved by people's governments of provinces (autonomous regions and municipalities) and certified by competent forestry administrations and the State Council have the authority to deploy temporary checkpoints for fire prevention. These checkpoints are responsible for fire inspection of all vehicles and personnel entering the fire zone to make use all motor vehicles are equipped with necessary firefighting equipment and tools as required. If PLA and CAPF have to enter the fire zone for responding to emergencies and performing other urgent tasks, they shall obtain approval from superior departments in charge and take necessary fire precautions.

No access to the fire zone during fire season is allowed without approval from competent people's governments at county level and above. And all vehicles and personnel entering the fire zone shall carry out activities according to the approved time, place and range under the supervision by competent forestry administrations of people's governments at county level and above.

3. Fire Safety Control System

Use of fire in the fields in any forest fire zone is prohibited during fire seasons. Where a fire has to be started in the fields for the purpose of pest, mice and frost control and in other exceptional circumstances, it is required to obtain approval from competent people's governments at county level and above and to take fire precautions as required. Live-fire drills and blasting in the fire zone shall be approved by the competent forestry administrations of people's governments at province (autonomous region and municipality) level, with necessary fire precautions in place.

As a general rule, the following "10–Nos" must be adhered to at all times: no shifting cultivation; no incineration & manure collection; no ridges & pool weeds burning; no pasture burning; no cigarette butt, match stick or other tinder; no firing for keeping warm or cookout; no torch; no firecracker setting-off, paper burning or candle lighting; no mountain burning for expelling wild beasts or hunting with firearms; no blasting or live firing.

4. Local Rules and Regulations on Fire Prevention

Local rules and regulations on fire prevention are the codes of conduct governing fire prevention as may be drawn up and executed by the civilians themselves living in the forest areas in South China. They constitute a supplement to laws and play a special and crucial role in forest fire prevention. Local rules and regulations have the following four characteristics in fire prevention practices:

(1) They address some problems not covered in national laws and regulations.
(2) They are drafted depending on local actualities, guaranteeing strong pertinence.
(3) They are easily accepted by the mass, guaranteeing mass involvement.
(4) The clauses are definite and concrete, guaranteeing operability.

5. Fire Reporting System

Forest fire prevention commands at different levels should always have staff on duty by turns in fire season. In addition to periodic work reporting to superiors, the staff should report promptly in case of fire.

6. Patrolling and Lookout System

In order to detect a fire timely in fire season, specialists should be assigned for patrolling, lookout and observation. Watchtowers should have specialists on duty day and night to monitor forest fires within the viewing area.

7. Cooperative Prevention System

In the neighboring forest segments of each two provinces, counties, villages (towns) and forestry administrations, it is imperative to develop and deploy a cooperative fire prevention system and organization, formulate cooperative fire prevention treaties and convene meetings on cooperative fire control.

8. Reward and Punishment System

The *Forest Law of the People's Republic of China* specifies that the units and individuals that contribute to forest fire prevention shall be rewarded timely; those violating forest fire prevention regulations or causing forest fires shall be investigated and punished timely; and the liabilities of leaders in charge shall be ascertained in case of serious forest fires.

Ⅳ. Ignition Source Control

1. Preparation of Ignition Source Distribution Maps and Forest Fire Incidence Maps

Ignition Source Distribution Maps should be prepared specificly to a forestry administration, forestry station or woodland with a specific area based on the statistics of forest fires recorded over 10 or 20 years in the forest area. The likelihood of ignition source can be classified into five levels marked by five different colors: Level 1 (red), Level 2 (light red), Level 3 (light yellow), Level 4 (yellow) and Level 5 (green). Forest Fire Incidence Maps and Ignition Source Analysis Charts should be revised or re-prepared at an interval of 5–10 years.

2. Delineation of Ignition Source Control Zones

(1) Basis for ignition source control zoning
Four major considerations:
① Ignition source type and quantity;
② Traffic conditions and terrain complexity;
③ Villages and rural settlements distribution characteristics;
④ Fuel type and combustibility.

(2) Classification of ignition source control zones
Three major types:

① Type I: the ignition sources are complex, and the quantity and incidence of ignition sources exceed the average of total ignition sources in the area; the area is characterized by underdeveloped traffic and complex terrain, and the forests susceptible to fire take up a large share of total forests; villages and rural settlements are scattered, and there is a great difficulty in ignition source control.

② Type Ⅱ: the ignition sources come in many types, and their quantity is equivalent to the average of total ignition sources in the area; the area is characterized by fairly developed traffic and relatively complex terrain; villages and rural settlements are relatively concentrated; and there is less difficulty in ignition source control.

③ Type Ⅲ: the ignition sources are simple, and their quantity is lower than the average of total ignition sources in the area; the area is characterized by developed traffic and simple terrain; forests are least prone to fire; villages and rural settlements are concentrated; and there is no difficulty in ignition source control.

(3) Ignition source control zoning units: zoning by forestry station or township or by county or forestry administration.

(4) Ignition source types: seasonal, perennial, mobile and key ignition sources.

3. Ignition Source MBO

Management by Objectives (MBO) is an effective modern economic management method. The method proves significantly effective when applied in ignition source control. Specifically, set an overall objective (e.g. decrement of total ignition sources in a forest) for ignition source control based on sufficient survey statistics, set the specific objectives for forest fire incidences in different ignition source control zones and for different types of ignition source, develop corresponding management measures against different management objectives to make sure administrative staff at all levels can identify their respective job objectives and duties, prepare their respective management plans and take effective measures to achieve the overall objective of ignition source control and forest fire prevention in a well-organized manner.

Ⅴ. Forest Fire Breaking

Forest fire breaking is the use of artificial and natural barriers to break a forest fire and stop spreading of the fire. Forest fire barrier systems should link in a grid with an area of 100–1,000 ha. The grid area may be reduced for planted forests and the forests in scenic zones, forest parks or natural reserves and increased for secondary and primary forests in distant mountains.

1. Firebreak

(1) Types

① Frontier firebreak: created on our side at the frontier with a width of 50–100 m to stop spreading of forest fires across the frontier.

② Railway firebreak: created along the sides of a railway with a width of 50–100 m to barrier on-board and artificial ignition sources that may induce a forest fire.

③ Boarder firebreak: created at the positions where forests join converge with farmlands, grasslands and settlements; its width varies with local topographic, vegetation and weather conditions, mostly 10–15 m wide in South China and 30–100 m in North China.

④ Other firebreaks: 50–100 m wide firebreaks created in the vicinity of settlements, forest farms and warehouses.

(2) Principles of Firebreak Construction

① The distribution of existing natural or artificial barriers must be considered to minimize investment;

② Primary firebreaks must stretch in the direction perpendicular to the direction of prevailing wind blowing in fire season of local forests, and 80–100 degrees is the preferred angle;

③ Firebreaks should be created on ridges or other lands with gently undulating terrain, few vegetation and poor soil as far as possible, without being in perpendicular to contour line.

④ The major tasks of creating a roadside firebreak include road widening and removal of bushes, weeds and litter.

(3) Construction methods: tractor-ploughing, mowing, chemical weeding and burning.

2. Fire-proof Roads in Forest

Fire-proof roads are traffic lines and also fire barrier belts in the forest. Road density in the forest is the foundation for mechanical firefighting and quick response and also a measure of silviculture performance in a region. The forest road density in developed countries is mostly higher than 10 m/ha, such as 17 m/ha in Japan, dozens of meters per hectare in Germany and 170 m/ha in some states in the USA. Up to 2014, the road density of key forests in China is merely 2 m/ha, which is much lower than that in developed countries.

Survey and design of fire-proof roads in a forest should be conducted in compliance with the *Technical Specifications for Forest Road Constructions* (LY5104), *Specifications for Survey of Forest Road Constructions* (LYJ115), *Specifications for Design of Forest Road Routes* (LYJ113), (LYJ114), *Specifications for Pavement Design of Forest Roads* (LYJ131), *Specifications for Design of Forest Roads, Bridges and Culverts* (LYJ106) and other applicable regulations.

3. Raw Soil Belt and Fire Trench

Raw soil belt is a ploughed soil belt that is typically created on high-value forest segments or in the vicinity of garages, workshops, forest depots and charking yards in the forest. It is generally 2 m wide and may be widened as needed on forest segments with high and dense weeds.

Fire trench is created to block ground fires. Besides, it can be used to stop spreading of low-intensity surface fires. Fire trenches are typically 1 m wide and 0.3 m deep and only created on forest segments with peat or thick humus layers.

Ⅵ. Green Fire Prevention

1. Green Fire Prevention

Green fire prevention is the use of green plants (including trees, bushes, vegetation and cultivated plants) for silviculture, afforestation, after-culture and cultivation to mitigate accumulation of forest combustibles, change the fire environment and enhance forest stands' flame retardancy and fire resistance so as to block or stop spreading of forest fires. It is a fire prevention method using green plants to reduce forest fire incidences, block or stop spreading of fires by employing various silviculture approaches.

Green fire prevention measures frequently used in production mainly include construction of fire belts and biological fire belts, stand improvement and fire prevention by afforestation.

2. Characteristics of Green Fire Prevention

(1) Enhance biodiversity and forest's disaster resistant ability.

(2) Maintain the forest's ecological balance, and produce fast-growing, quality and effective forests with minimal disaster incidences.

(3) Promote environmental protection and territorial management.

(4) Permanent facilities for forest fire prevention.

(5) Small investment, quick returns and high economic benefits.

(6) The planting density of fire belts should be determined based on the biological characteristics and structure of tree species, and should be greater than that of plantation forests. The plant spacing should be controlled between 1 m × 1 m and 2 m × 2 m.

(7) Fire belts should have compact structure, extending both horizontally and vertically. A multi-layer structure should be adopted, forming multi-storied forest belts consisting of large and small trees mixed with shrubs.

(8) Tree species with strong fire resistance and good adaptability to local conditions should be planted in the fire belts, and should meet the following requirements:

① Dense foliage, high moisture content, strong fire resistance, low oil content and difficult to burn;

② Rapid growth, quick crown closure, strong adaptability and good sprouting ability;

③ Good shade tolerance of understory trees and good interspecific relationship with canopy trees;

④ Free from diseases, parasites and pests.

(9) Tree species for fire belts should be suitable to local conditions. For creating fire belts in northern China, arbor species such as mandshurica ash tree, manchurian walnut tree, amur cork tree, poplar, willow, lime tree, elm, maple, bird cherry and larch should be planted, while shrub species such as honeysuckle, winged spindle tree, elder and white lilac should be planted. For creating fire belts in southern China, arbor species such as schima superba, holly, corylus chinensis franchet, michelia macclurei, acacia auriculiformis, cork oak, daphniphyllum, coral tree, schisandra propinqua, clerodendrum inerme, broussonetia kazinoki, paper mulberry, castanopsis kawakamii, machilus thunbergii, metrosideros excelsa, camellia oleifera, alder, syzygium cumini, chenopodium, castanopsis fissa, myrica rubra, cyclobalanopsis glauca, and podocarpus nagi should be planted, while shrub species such as camellia oleifera, schefflera octophylla, eurya japonica, psychotria rubra, tea tree (Yunwu tea) should be planted.

3. Forest Fire Prevention by Afforestation

Forest fire prevention by afforestation refers to a set of green fire control measures for reducing forest combustibility and preventing forest fires through afforestation, afforestation, logging and other measures in the process of forest afforestation, in order to adjust the flammable components of forest stand, adjust the stand structure and enhance the fire resistance of the forest. Key measures of forest fire prevention by afforestation are as follows:

(1) Create mixed conifer-broad-leaved forest, improving fire resistance of plantation forests.
(2) Improve and make use of natural broad-leaved forests.
(3) Improve forest stand.
(4) Carry out inter-planting on forest land.
(5) Take afforestation and management measures.
(6) Reduce fuel buildup through biological control measures.

Ⅶ. Black Fire Prevention

1. Black Fire Prevention

Black fire prevention is the general term of burning methods for preventing forest fires by prescribed burning of combustibles, in order to reduce fuel buildup and forest combustibility through or by creating firebreaks through burning. Prescribed burning, also called controlled burning, is vividly called “black fire prevention” or “fight fire with fire” as the burned area is black.

2. Black Fire Prevention Measures

(1) Burning of firebreaks.
(2) Burning of ditches and meadows.
(3) Comprehensive spot burning.
(4) Prescribed burning inside the forest.
(5) Fight fire with fire.

3. Conditions for Safe Burning

(1) Season and time of burning: to ensure safe burning, burning should be carried out at

the right season and time. Burning of logging residue should be carried out in winter with heavy snow or in the plant growing season.

(2) Weather conditions for burning: burning should be carried out under the stable weather conditions. Half a day to 3 days, not more than 5 days following the precipitation is the right time for burning. The wind speed on the date of burning should be within force 3, not more than force 4. The relative humidity should be within 40%–60% with low temperature.

(3) Condition of combustibles for burning: combustibles in the burning area should be evenly distributed and the moisture content of combustibles should be within 15%–20%.

4. Common Igniter Types

Drip-feed igniter, type 17 igniter, 76-regulating portable igniter, BD igniter, SDP-1 portable adjustable flow igniter, SID portable booster igniter, BOD-1 backpack multi-purpose igniter and BCD-1 injection type igniter.

Ⅷ. Forest fire control plan

1. Forest fire control plan

Generally, a forest fire control plan is composed of three parts: main body, schedules and attached drawings. The main body includes preface, general introduction, analysis of situations and circumstances, general ideas, construction contents and tasks, main construction projects, investment calculation and financing, guarantee measures, and main suggestions on policies, etc.

The schedules include tables statistically setting out the basic information of forest fire prevention and control, forest fires happened, investment estimation and so on. The attached drawings comprise forest resource distribution maps, fire danger rating and zoning maps, construction layouts, etc.

The general term for the forest fire control planning is 10 years, and can be divided into two stages: a short-term stage and a long-term stage.

2. Planning principles

(1) Forest fire control should be people-oriented and science-based: During the forest fire control, priority shall be given to the life safety and properties of people living in the forest areas and the fire crews and brigades working on the front lines of firefighting. Meticulous organization and scientific command shall be obtained to reduce the casualties and property losses caused by forest fires.

(2) Forest fire control efforts should be specific and practical: The forest fire control planning must be suitable for China's national conditions and forest conditions. Construction projects shall be selected according to the actual situations, and technical measures shall be scientific, economic and applicable.

(3) Forest fire control policies shall suit for different zones and give prominence to the key points: The forest area under control shall be zoned reasonably according to actual situations such as forest resources, fire danger ratings and fire occurrence and development rules, and applicable forest fire management and control measures should be taken accordingly. As for key construction regions and tasks, more efforts and investments shall be input.

(4) Forest fire control shall aim at seeking both temporary solutions and a permanent cure (especially the latter one): The infrastructures and equipment for the forest fire control shall be improved comprehensively, with focus on the construction of fundamental long-term projects such as emergency access roads and biological fire breaks. Supporting policies and guarantee measures shall be strengthened and the construction of long-term mechanism shall be highlighted.

(5) Forest fire control shall be led by science, technology, reform and innovation: The technical content of the forest fire control shall be increased continuously through the active development, introduction and promotion of advanced and practical forest firefighting machinery and technology. Market access, service purchase and other mechanisms shall be introduced to enhance the standardization and normalization of forest fire control.

(6) Forest fire control shall lay equal stress on construction and management by joint efforts in both aspects: The quality of construction projects shall be controlled strictly and the operation, maintenance and management of these projects shall be strengthened to ensure the full utilization of facilities and equipment resources, the continuing effectiveness of the construction results, and the maximization of the efficiency in fund utilization.

3. National forest fire control planning of China

(1) Construction zones: In light of the distribution of forest resources, climate, geography, forest fire danger ratings and other factors in China, and to meet the basic requirements such as forest fire control coordination convenience, scientific and reasonable layout, and clear construction emphases, we divide China into six construction zones for forest fire control purpose, including the Northeastern Zone, the Southwestern Zone, the Northwestern Zone, the Southeastern Zone, the Central China Zone, and the North China Zone. Among them, the Northeast Zone and the Southwest Zone are key construction zones.

(2) Overall goal: The overall goal is to keep the forest fire incidence at 1‰ or below by establishing three systems: forest fire prevention system, forest fire suppression system and forest fire support system. These three systems can comprehensively improve the equipment and infrastructures for forest fire control, enhance the abilities in early warning, monitoring, emergency response and fire suppression, and realize modern fire prevention and control procedures, standard fire management work, professional fire crews, and scientific fire suppression measures.

(3) Main construction contents: The publicity and education for forest fire control includes activities in education centers, advertisements on public transportation vehicles and outdoor publicity slogans and signs. The construction of the forest fire early-warning system mainly comprises fire danger factor monitoring stations and combustible factor collection station. The ignition source management covers patrol cars, site enforcement recorders, meters and instruments, distance-measuring equipment, IR residue fire detectors, GPS, personal protective equipment, centralized burning pools. The construction of the fire lookout and monitoring system mainly include ground patrol and lookout and monitoring systems. The construction of access roads and fire break systems mainly comprises fire command centers, communication and terminals, communication command vehicles. The

construction of firefighting machines and tools covers the machines and tools for firefighting, personal protection and wildness survival, engineering machinery and fireproofing vehicles. The construction of aerial firefighting projects mainly includes the helicopter landing sites in central stations and forests, firefighting pools in forests and firefighting equipment. The camp project for professional fire brigades mainly includes the construction of architectures and training venues. Other infrastructure projects mainly include fireproof material warehouse, checkpoints and fire training bases.

In addition, the national forest fire control planning shall also cover the construction of forest fire brigades, scientific researches, training, fire case reconnaissance, fire loss assessment and so on.

CHAPTER Ⅲ

Forest Fire Forecast and Communication

Ⅰ. Forest fire monitoring

Forest fire monitoring is primarily conducted at four levels: ground patrol, monitoring from lookout tower (or video), air patrol and satellite monitoring.

1. Ground patrol

(1) Main tasks: conduct publicity of forest fire control, strictly control ignition sources, eliminate fire hazards, timely detect fires, rapidly report fires, actively suppress the fires, and realize comprehensive monitoring with the aids of lookouts.

(2) Organization mode: the ground patrol team is generally composed of foresters, stationary forest police squads, motorcycle patrolmen and water patrolmen who patrol along a certain route.

2. Main tasks of monitoring from lookout towers (or video)

The main tasks are to monitor fire hazards, watch forest fires, identify places where fires occur, and report fire in time. By the end of 2013, the coverage rate of fire monitoring from lookout towers in key forest areas nationwide has reached 68.1%.

3. Construction requirements of lookout towers

(1) Requirements on site selection of lookout towers

The lookout towers shall be:

① On a raised peak or highland at high altitudes;

② With wide vision and high visibility;

③ Free from hazards arising out of other interferences or natural disasters;

④ Close to the settlements and roads.

(2) The setting density of the lookout towers (or video) shall be determined according to the terrain, topography, forest distribution, observation methods, visibility and other conditions. Through visual observation and simple instruments (such as telescope), the lookout radius is generally 10–20 km. If the video system and other instruments are used, the observation radius is dependent on the performance of such systems and instruments.

(3) The sight lines of the lookout towers (or videos) shall be interleaved and crossed, and the repeated observation part shall be no less than 1/5 of the overall observation area. There should be no blind spots in the group control area of the lookout towers (or videos).

(4) The lookout towers shall be of steel structure, brick concrete structure or reinforced concrete structure.

① If the steel structure is adopted, a lookout tower shall be consist of a tower foundation, a tower base, a tower support, a tower cupola, a lifting system (stairs or elevator), a weight system, a security system, a lightning protection system, etc.

② If the brick concrete structure or reinforced concrete structure is adopted, a lookout tower shall be consist of a tower foundation, a tower base, a tower cupola, an ascending and descending system (stairs, step platforms, stair balustrade, etc.), a security system (rails, handrails, etc.), a lightening protection system, etc.

(5) According to actual conditions, the lookout tower may choose square or angle truss or may be built into a rectangular (cylindrical) or pyramidal (conical) shape. The lifting system shall be an inner-stair lifting system or an automatic lifting system, and the stairs shall be lifted in a folded way.

(6) The height of the lookout tower shall be determined according to the topography, the

height of the surrounding forests and the scope of control. In low-relief areas, the cupola of the lookout tower shall be higher than the highest crowns of surrounding trees, and the height difference between the cupola floor and the highest crowns of surrounding trees shall be no less than 2 m. In hilly areas, the height of the lookout tower shall be 10–26 m. As for raised mountain peaks and places with clear line of sight, no tower support or tower body but a tower cupola may be required. The lookout towers in young-and-middle-aged forests shall be erected at a height based on the height of mature forests.

(7) A lookout tower should include staff accommodation rooms, warehouses and other facilities. If such facilities and the lookout tower are built separately, they shall not be far from each other, and the distance between them shall not be more than 100 m.

(8) The lookout tower may be erected where the lookout height is no more than 10 m above the ground. A two-story brick and concrete structure may be selected for such lookout tower, with the upper story being the tower cupola and the lower story being the ancillary facilities.

(9) The lookout towers aiming at omnibearing observation shall be set in the center of a forest area or a control area, and those aiming at single-direction observation shall avoid observation against the sun.

(10) The lookout tower shall be far from the high-voltage transmission lines, and the shortest distance shall not be less than 100 m.

(11) The lookout tower must be equipped with a lightning protection device which is technically reliable, safe and applicable. The maximum earthing resistance shall not exceed 10 ohms, or otherwise, safety measures and corrective regime shall be put forward.

(12) The grounding lead of the lightning protection device shall be closely connected to the lightning rod. No detachment or breakage is allowed. An additional length about 0.5 meter shall be reserved on the grounding lead.

(13) Clear signs shall be marked on the ground for lightning and grounding devices. There shall be a 0.8–1 m thick cover soil layer above these devices. The ground surface after be-

ing covered with soil shall not be lower than the surrounding ground.

(14) The lookout tower shall be equipped with living equipment, observing equipment, positioning equipment, communication equipment, etc.

(15) The video monitoring system for fire prevention shall be arranged in key places according to the comprehensive layout of the lookout towers. The video monitoring system mainly includes information acquisition equipment, data transmission equipment, image processing and analysis equipment and ground monitoring station.

4. Air patrol

China's aviation forest protection tasks are mainly undertaken by two departments: Northern Aerial Forest Protection Station and Southern Aerial Forest Protection Station. These two stations function in aerial forest fire control, forest fire control coordination, and fire control material storage, and undertake the aerial forest protection masks such as air patrol, forest fire report, forest fire reconnaissance, aerial command, extinguishing operation with helicopters, aeroplane fleet, fast roping (slipping) from helicopters and hanging buckets (bags) from helicopters, aerial distribution of fire prevention ads, emergency aid on fire field, and on-field investigation.

Up to 2014, the Northern Aerial Forest Protection Station has 16 aerial forest protection stations which are located respectively in: Heihe, Nenjiang, Jiagedaqi, Tahe, Yichun, Mudanjiang, Jiamusi, Xingfu and Dongfanghong in Heilongjiang Province; Genhe, Hailar, Zalantun, Ulanhot and Mangui in the Inner Mongolia Autonomous Region; and, Dunhua and Changbai Mountain in Jilin. There are 48 air patrol airlines with a total length of 18,700 km.

Under the direct jurisdiction of Southern Aerial Forest Protection Station by 2014, sit aerial forest protection stations are set respectively in Pu'er, Baoshan, Lijiang, Chengdu, Xichang and Baise. In these jurisdictions, there are 12 seasonal aerial forest protection bases which are located respectively in Kunming, Dali, Daohu County, Panzhihua, Mianyang, Liuzhou, Wuzhou, Guiyang, Ganzhou, Luoyang, Qingyuan and Meizhou. The

Southern Aerial Forest Protection Station is responsible for managing the aerial forest protection businesses in 18 provinces (autonomous regions and municipalities) including Yunnan, Sichuan, Chongqing, Guizhou, Tibet, Guangxi, Guangdong, Jiangxi, Henan, Hubei, Hunan, Shanghai, Jiangsu, Zhejiang, Anhui, Fujian, Shandong and Hainan. So far, it has conducted the aerial forest protection businesses in 8 provinces (autonomous regions and municipalities) including Yunnan, Sichuan, Chongqing, Guizhou, Guangxi, Guangdong, Jiangxi and Henan. These aerial stations (bases) shares 158 airlines with a total length of 75,852 km.

Up to 2014, China has preliminarily established an aerial forest fire control system with Northern Aerial Forest Protection Station and Southern Aerial Forest Protection Station as the core and secondary aerial forest fire control stations as backbones. Every year, 80 aircrafts are hired to complete the 7,000-hour air patrol task and the patrol area reached 3,110,000 km^2. The application of large firefighting aircrafts such as Mi-26, Ka-32 and Mi-171 make the aerial forest fire prevention become one of the major approaches means to extinguish a forest fire under complicated geographical conditions.

5. Satellite monitoring

(1) Main functions: detect fire, monitor fire behavior, evaluate post-fire loss, draw distribution map of inflammable in forest area and other professional fire prevention maps, carry out forest fire danger rating calculation, conduct measurements, transmit weather factors, and so on.

(2) Main types of satellites: polar orbiting meteorological satellites, land-resource satellites, geosynchronous earth orbit satellites, low earth orbit satellites and the like. For example, the NOAA satellite is a double-satellite system consisting of an AM-orbit satellite and a PM-orbit satellite, and the system can run around the earth for at least 4 times in 24 hours. Russia uses NOMOS, a low-orbit satellite system, to monitor forest fires. The NOMOS consisting of 6–8 low-orbit satellites (400 km above the ground), with an average observation interval being 1 hour, can detect forest fire areas smaller than 0.01 km^2, and its precision in locating ignition points is within 0.3–0.5 km.

(3) The Forest Fire Satellite Monitoring Center of State Forestry Administration is in Bei-

jing, governing 3 secondary satellite forest fire monitoring centers located in Harbin, Kunming and Urumqi respectively. This center can monitor the orbits of FY3, EOS-MODIS and NOAA series satellites, handle over 13,000 orbits every year, generate more than 4,500 images, and report 12,000 hot spots. With many blind spots and a spatial resolution of 1 km, the forest fire satellite monitoring in China is still unable to form a full coverage nationwide.

Ⅱ. Forest fire forecast

1. Types

Forest fire forecast is generally divided into fire weather forecast, fire occurrence forecast and fire behavior forecast.

Fire weather forecast basically provides weather forecast for forest fire dangers based on weather factors. Generally, rather than the ignition source condition and the fire behavior, only the possibility of a forest fire caused by the weather conditions is forecasted by giving full consideration to temperature, relative humidity, precipitation, wind speed and continuous drought days.

The forest fire forecast is to forecast the possibility of a forest fire with full consideration to weather factors, combustible conditions (dryness, wetness, load, flammability) and ignition sources (types and distribution and frequency). It includes the possibility of lightning-caused fires, artificial fires and other fires.

Forest fire behavior forecast is to predict forest fire behaviors after the occurrence of the fire such as spread speed and fire intensity, based on the weather, combustible conditions and topographic factors (slope, slope exposure, slope position and altitude).

2. Types of fire weather forecast

Forecasts can be classified into ultra-short-term forecast (within 6 hours), short-term forecast (6 to 12 hours), short-term forecast (12 to 72 hours), medium-term forecast (3 to 10 days), long-term forecast (10 days to 1 quarter) and ultra-long-term forecast (more than 1 year). In contrast with weather forecasts, there are only short-term and medium-term fire weather forecasts.

3. Main fire weather forecast methods

Fire weather forecast can be conducted with different methods including the burnt ground-cover, wooden cylinder, pine and cypress branch, comprehensive index, wind speed supplement plus comprehensive index, actual effect humidity, fire danger scale, fire index, spread index, etc.

Ⅲ. Communication of forest fire control

1. Types of communication of forest fire control

According to the transmission lines and media, the communication can be divided into 2 categories: wired communication and wireless communication.

The wired communication refers to a communication mode which transmits information through wires, cables and optical fibers, and it is a major communication way for the forest fire control work in China, such as wired telephone.

The wireless communication refers to a communication mode which uses radio waves to transmit signals in space. The wireless communication of the forest fire control mainly includes short-wave communication, ultra-short wave communication and microwave communication.

According to the channel relay modes, the communication can also be divided into satellite communication, balloon communication, ground relay communication and so on. Based on the type of information transmitted, the communication can be divided into voice communication, digital communication, image communication, etc.

2. Communication network of forest fire control

The communication network for forest fire control is a communication network which consists of communication nodes built in different forest points with exiting wired and wireless communication modes. It is used to transfer information on forest fire control, give fire alarm and complete other tasks. China's communication network for forest fire control can be divided into four tiers according to the relevant management systems, jurisdictions and duties.

(1) Tier 1 Network: with the national forest fire prevention command center as the master station, and provincial forest fire prevention command centers as subordinate stations.

(2) Tier 2 network: with provincial forest fire prevention command centers as the master station, and prefecture-level (cities, districts, leagues) forest fire prevention command centers as subordinate stations.

(3) Tier 3 network: with prefecture-level (cities, districts, leagues) forest fire prevention command centers as the master station, and the county-level (cities, districts, flags) forest fire prevention command centers as subordinate stations.

(4) Tier 4 network: with county-level forest fire prevention command centers as the master station, and the basic units under the jurisdiction of prefecture-level (cities, districts, leagues) forest fire prevention command centers as subordinate stations.

Temporary communication network shall be established according to actual needs, in order to communicate with the fire command centers directly or through information transfer. A reasonable networking scheme shall be selected to realize smooth ground-to-ground, air-to-air, and ground-to-air information transmission.

Up to 2014, the coverage rate of China's communication network for forest fire control is only 67.7%. There are many communication networks, but they cannot be fully interconnected.

3. Settings mode of radio stations for fire control

The said setting modes include direct communication, indirect communication, medium-long distance composite communication, reversible communication and multi-relay communication.

CHAPTER Ⅳ

Forest Fire Suppression

Ⅰ. Principles and procedures of fire suppression

1. Fire extinguishing principles

Three basic elements are needed for a forest fire: Forest combustible, oxidizer (air, oxygen) and ignition source (at a certain temperature). If any of these three elements does not exist, the combustion stops. Therefore, a fire can be extinguished as long as any of these three elements is damaged or controlled.

2. Basic extinguishing methods

The basic extinguishing methods include: Smothering (isolating the combustible from air), cooling (reducing the temperature) and isolating (sealing the combustible).

3. Basic extinguishing principles

There are three principles: Extinguishing a fire in early stage, extinguishing a fire when it is small, and extinguishing a fire till it completely goes out.

4. "Early, Fast, Prompt"

(1) Early: Early preparation, early detection and early dispatch.
(2) Fast: Leaders shall act fast, and extinguishing measures shall be taken fast.
(3) Prompt: No small fire is allowed to become an overnight one. In other words, small and incipient fires shall be extinguished before 8:00 AM of the next day.

5. Firefighting procedures and stages

(1) Firefighting procedure: A. control; B. extinguish; C. consolidate.

(2) Five stages of firefighting process: Initial extinguishing stage, open fire extinguishing stage, ember extinguishing stage, fire field guarding stage, and evacuating stage.

Ⅱ. Firefighting organization

1. Command system for firefighting organization

The command system for firefighting organization consists of standing forest fire control command organizations established by governments at all levels and temporary forest fire control command organizations.

2. Five levels of command organization for forest fire control

The five levels include: Forest fire prevention office under State Forestry Administration, provincial (autonomous regions and municipalities) forest fire prevention command centers, prefecture-level (cities, districts, leagues) forest fire prevention command centers, county-level (cities, districts, flags) forest fire prevention command centers, and town-level (towns, fields and offices) forest fire prevention command centers.

By 2013, there had been 3,342 national command organizations for forest fire control and 22,000 full-time forest fire control officers. Nine provinces (autonomous regions) have established a full-time commander mechanism, and ten provinces (autonomous regions) have established early warning and monitoring organizations.

3. Command system for temporary firefighting organizations

The command system for temporary firefighting organizations comes into being with the emergence of fire scene. The temporary firefighting organization and command system, which dissolves upon the completion of the firefighting tasks, generally has 3 organization forms.

(1) One-person independent command center: It is applicable when commands are given to 1–2 firefighting teams on the same fire field. The prerequisites often include: Measures can be taken immediately after a fire is detected; the fire can be controlled within 3 hours; open fire can be extinguished within 5 hours, and the area of the fire field should be smaller than 100 ha.

(2) Single-layer command center: Such command center usually consists of 3 to 5 people, including one commander and several deputy commanders. The prerequisites often include: Commands are given directly to front-line firefighters, and the area of the fire field should be smaller than 100–1,000 ha.

(3) Multi-layer command center: The multi-layer command mode is applicable to extinguishing forest fires covering an area larger than 1,000 ha, or cross-town, cross-town, cross-city, major and serious forest fires. In case of more than one independent fire scene, many firefighting teams and complex conditions, a multi-layer command center is also needed.

4. Five major firefighting teams

They are professional fire brigades, semi-professional fire brigades, aerial ranger teams, armed forest forces and community-based volunteer fire brigades.

Sticking to the principles of "affordable, supportable, practical" and in order to meet the general requirements on "diversified forms, integrated command, standardized management, normalized equipment, regular training and scientific operation", China has strengthened the construction of forest fire brigades and basically established an integrated firefighting system, significantly improving the overall fighting capacity. The established firefighting system takes local forest fire brigades as the main force, the armed forest forces as the assault force, the aerial ranger teams as the offensive force, and the Liberation Army, the armed forces, reserve forces, militia emergency detachments, forestry workers and the masses as the backup force.

By the end of 2013, there have been 3,264 professional forest fire brigades composed of more than 110,000 people, 26,000 semi-professional forest fire brigades composed of

640,000 people. There have been 12 national material warehouses established in cities such as Beijing, Yakeshi, Harbin, Kunming and Changsha, 39 provincial material warehouses and 1,435 county-level material warehouses.

5. Frontline command center and commander for forest fire control

During the forest fire suppression, a front-line or temporary command center needs to be established to give orderly commands to strengthen the leadership of the forest fire control work. A primary leader having actual firefighting experience shall be elected as the commander of the fire field. The location of such command center shall meet the following five requirements: a. Close to the fire field and convenient for command work; b. Helpful to the understanding of fire situation; c. Facilitating the gathering and calling of fire brigades; d. Easy to connect and communicate; e. Safe. Selecting safe location which should be marked with a red or colorful flag.

The commander is the key to the forest fire control work. All firefighting plans and schemes shall be established and implemented by the commander. The quality and capacity of the commander will decide the success or failure of the fire extinguishing work. A good commander shall have the basic skills in seven aspects: Observation, judgment, decision making, response, expression, communication and command.

At the same time, a good commander shall have five professional competences: a. Master the firefighting knowledge; b. Quickly grasp fire situation and predict the change of sire situation; c. Promptly take corresponding measures; d. Practice and implement the firefighting scheme through the members of forest fire brigades under control; e. Keep calm and decisive in all cases and always lead and command own brigades in a proper way; f. Strike a proper balance between work and rest of fire brigades to maintain their vigorous morale.

Ⅲ. Extinguishing materials and equipment

1. Extinguishing materials and equipment

Extinguishing materials and equipment include fire extinguishing equipment, personal protection equipment, wilderness survival equipment, transportation equipment, engineering machinery and equipment, communication and command equipment, and spare equipment.

(1) Hand tools for extinguishing: Mainly including No. 1 tools, No. 2 tools, axes (machetes) shovels, hammers, handsaws and a fire clearing package.

(2) Extinguishing machines and tools: Mainly including pneumatic extinguisher, pneumatic water-based extinguisher, firefighting water gun, mobile pump fire extinguishing system, chain saw, brush cutter, extinguisher oiler and igniter, etc.

(3) Personal protective equipments: Mainly including protective helmet, intercom helmet, protective glasses, anti-noise ear-muffs, flame retardant clothing, fireproofing gloves, anti-stab shoes, water/fire/scrap-proofing boots and fireproofing hood.

(4) Wilderness survival equipments: Mainly including portable tents, down sleeping bags, moisture-proof mattress, air bed, outdoor cooking utensils, field rations, first-aid kit, medicine box and directional lamp.

(5) Transportation equipments: Mainly including troop carriers, water tender, cooking vehicles and command vehicle, etc.

(6) Engineering and mechanical equipments: Mainly including bulldozers, excavators, ditching machines, etc.

(7) Command and communication equipments: Including handheld interphones, vehicle-mounted station, mobile relay station, air station, command tent, emergency light, forest fire plotter, GPS, handheld meteorograph, portable projector, fire field reconnaissance equipment box and the command map etc.

(8) Spare equipments: Mainly including solar power supply, generator, manual motor, etc.

2. Introduction of main fire extinguishing machines and tools

(1) No. 1 and No. 2 tools: During the extinguishing process, we often tie tree branches together to form brooms, or tie wet sacks to wooden sticks. These tools are called "No. 1 tools". No. 2 tools are improvements based on No. 1 tools: Strip the outer layer of the waste or used car tires; cut the inner layer into rubber strips which are 2–3 cm wide and 80–100 cm long; and use nails or steel wires to fix these rubber strips onto the hard plastic tube or wood stick with a diameter of about 3 cm. The fixed rubber strips look like a broom. Such tools are mainly used to extinguish weak surface fires.

(2) Igniter: It is mainly used to set up fire breaks, ignite back fire, shorten the fire line (change the inside fire into outside fire), ignite for self rescue, and burn plans. During the forest fire control, drop-type igniter and gun-type igniter are often adopted.

(3) Firefighting water gun: It mainly consists of two parts: A bag (or plastic barrel) and a water gun. The bag (or plastic bucket) is a container used to hold water or chemical extinguishing agents. The firefighting water gun can be carried on back with straps. The water gun (hand pump) consists of a pump barrel, a stopper rod, a water inlet valve, a water outlet valve and a nozzle. This tool is mainly used to extinguish initial forest fire, suppress weak surface fire, clear fire field and water fire breaks. It can be used in combination with pneumatic extinguishers and No.2 tools.

(4) Dry powder fire extinguisher: Adopting dry powder of sodium salts as extinguishing agent, it detonates an appropriate amount of explosives with an ignition device so that the dry powder is thrown into the space to form powder fog which has an extinguishing effect. This tool can be used to put out initial surface fire, block fire head, suppress fire, and create conditions allowing firefighters to approach the fire lines.

(5) Axe, shovel, hatchet and extinguishing rake: Axe is mainly used for cleaning shrubs, cutting down small threes and blow-downs, setting fire breaks, erecting camps, and building temporary helicopter landing sites; the shovel is mainly used for cutting roots of trees and grasses, isolating ignition sources, extinguishing embers and hidden fires, and digging ditches to put out underground fire; hatchet is mainly used for cutting down branches, shrubs and weeds to facilitate the access of firefighters; extinguishing rakes can be divided into coarse-

tooth and fine-tooth rakes and mainly used to clean away combustibles when setting fire breaks.

(6) Extinguishing fuse: This tool is composed of storage and transportation device, a detonator, an igniter and a fire extinguishing rope. It is mainly used at the time of setting fire breaks and putting out surface and crown fires.

(7) Chain saw and brush cutter: Chain saw is mainly used to set permanent and temporary fire breaks, and brush cutter is mainly used to clean combustibles on forest ground, as well as flammables and combustibles on the fire break.

(8) Pneumatic extinguisher: Pneumatic extinguisher, using the fan to produce a strong wind, can blow away the heat produced by the burning of combustibles and cut off the connection among burned, burning and unburned combustibles, thus extinguishing the fire. Pneumatic extinguisher mainly consists of a gasoline engine, a self-suction centrifugal fan, a front handle, a back handle, a strap and other accessories. The fan is composed of an impeller, a wind tube and so on. The impeller is driven by the engine and rotates at high speed to produce a strong wind to extinguish the fire directly.

During the fire season, the engine and fuel tank of pneumatic extinguisher should be inspected every day. The place where the pneumatic extinguisher is stored should be ventilated, dry and away from fire. In the non-fire season, the fuel in the fuel tank shall be emptied before the storage of the pneumatic extinguisher. After being wiped clean, the pneumatic extinguisher shall be stored, and regularly checked and maintained.

(9) Firefighting water pump: The fire extinguishing pump is Tier Ⅱ micro centrifugal pump which is designed for forest fire control and features high lift and low flow. The working principle of this firefighting water pump is that: Water cools down combustibles; when heated, water evaporates and produces a large amount of vapor to take up the combustion space, stopping the air from entering into the combustion area and reducing the oxygen content in local part of the combustion area. At the same time, after being pressurized by the pump, water can strike the combustibles with higher kinetic energy, thus dispersing combustibles and reducing combustion intensity. Extinguishing by sprinkling water is the most effective way to put out a forest fire, especially a crown fire and an un-

derground fire.

(10) Amphibious forest firefighting truck: It is a modification of 63-Type Tracked Armored Vehicle. This truck can run in the conditions of forest, grass pond and various abominable mountain areas at a high speed, thus being suitable for firefighting operations in forest areas and forests.

(11) J-50 Airborne Forest Firefighting Vehicle: J-50 airborne forest firefighting vehicle consists of a J-50 vehicle body, a special water tank and a fire-extinguishing water pump. The water tank can hold 2.5 tons of water. The vehicle is a large extinguishing machine, and it is separable and combinable, and can be used for multiple purposes.

(12) "531" firefighting vehicle: It adopts a "531" armored vehicle as carrier, and use Canadian MARK water pump as water suction and injection pumps. The water container consists of two parts: One is inside the vehicle, the other is on the vehicle. "531" firefighting vehicle has the advantages of simple operation, convenient use and high extinguishing performance. The vehicle is capable of carrying 21 tons of water at most. Its minimum water absorption time is 5 minutes, maximum water spraying distance is 30 m, and optimal water spraying distance ranges from 8 to 15 m.

(13) Aircraft: The aircraft for air extinguishing is mainly composed of an engine, a body, wings, a tail and a landing gear. According to the shape of the body, the aircraft can be divided into fixed-wing aircrafts and non-fixed-wing aircrafts (i.e., helicopters). The main task of fixed wing aircraft is to patrol, alarm, detect fires, distribute fire control ads, transport firefighting materials, conduct chemical extinguishing procedure, and train investigators. In addition to transporting firefighters, the helicopter for forest fire control shall conduct air patrol, extinguishing by landing firefighter with ropes, extinguishing by sprinkling, and other emergency tasks. Due to the short flight and high charge of helicopter, the arrangement of helicopters for air patrol shall address the key problems, and the best location, best patrol time and best route shall be selected, thus ensuring a scientific and effective flight.

Ⅳ. Tactical countermeasures during firefighting

1. Basic concepts

(1) Extinguishing tactics refers to the skills of successfully extinguishing a forest fire by organizing and implementing forest fire control work with the most reasonable and effective methods, correctly using various firefighting machines and tools, and giving full play to the capabilities of all firefighting forces.

The main contents of the extinguishing tactics include: Basic extinguishing principles, team calls, coordinated action, commands, extinguishing actions, logistical and technical support, etc.

(2) Classification of tactics

① Classification by basic extinguishing methods: Direct extinguishing tactics and indirect extinguishing tactics;

② Classification by types of fire brigades: Aerial extinguishing tactics, ground extinguishing tactics and cooperative firefighting tactics;

③ Classification by the extinguishing scale: Tactics for individual extinguishing behavior, tactics for the extinguishing by single fire brigade, and tactics for the extinguishing by multi fire brigades.

(3) Firefighting tactical countermeasures: Refers to specific guidelines and principles of action adopted and employed during firefighting. The tactical countermeasures are guided and restricted by strategic countermeasures and have a direct impact on the realization of strategic countermeasures. Strategic countermeasures are achieved through tactical countermeasures.

During direct extinguishing process, it should be stressed that: Superior forces, machines and tools shall be gathered quickly in a certain time and place, to extinguish a fire in a rapid, effective and active way. During indirect extinguishing, favorable conditions such as terrain and surface features shall be fully used to set up a fire break quickly, thus sealing the forest fire. And then, ignite black burn fires, or take the initiative, or adopt other methods to extinguish the fire.

2. Basis for determining tactical countermeasures

Forest combustible, topographic conditions, meteorological factors, fire behavior, fire field situations and extinguishing abilities are all important basis for determining the tactical countermeasures for forest fire suppression.

(1) Forest combustible

① Type of forest combustion: Forest combustion can be divided into 3 types by the flammability of the dominant tree species (groups).

a. Difficult-flammable
Alnus spp.
Bambusoideae
Castanopsis spp. (including *C.eyrei, C. carlesii, C. sclerophylla*)
Cyclobalauopsis spp.
Fraxinus mandshurica Rupr.
Juglans mandshurica Maxim.
Paulownia spp.
Phellodendron amurense Rupr.
Phoebe spp.
Robinia pseudoacacia L.
Schima spp.
Mixed broadleaf forests (without obvious advantage)

b. Combustible
Abies spp.
Betulla spp.
Cryptomeria spp.
Cunninghamia lanceolata (Lamb.) Hook

Davidia involucrate Baill.

Larix spp.

Metasequoia glyptostroboides Hu et Cheng

Picea spp.

Populus spp.

Sassafras tsumu

Taxus spp.

Tilia spp.

Mixed conifer-broadleaf forests

Hard broad-leaf forests (*Acer mono* Maxim., *Fagus* spp., etc.)

Soft broad-leaf forests (*Pterocarya stenoptera* C.DC., *Salix* spp., *Acer* spp., *Catalpa* spp., *Casuarina Adans* L., *Melia* spp., etc.)

Miscellaneous woods

c. Flammable

Castanea spp.

Cinnamomum camphora (Linn.) Presl.

Cupressus funebris Endl.

Eucalyptus spp.

Keteleeria spp.

Liquidambar formosana Hance.

Lithocarpus spp.

Quercus spp. (*including Quercus dentata, etc.*)

Pinus armandi Franch.

Pinus densata Mast.

Pinus densiflora Sieb.et Zucc.

Pinus kesiya Royle ex Gordon var. langbianensis (A.Chev) Gaussen.

Pinus koraiensis Sieb. et Zucc.

Pinus massoniana Lamb.

Pinus sylvestris var. mongolica Litv.

Pinus tabulaeformis Carr.

Pinus thunbergii Parl.

Pinus yunnanensis Franch.

Mixed coniferous forest (without obvious advantage)

Shrubbery

② Distribution of combustibles: If combustibles are distributed horizontally, the amount, centralization, decentralization and distance of combustibles can affect the intensity and spread speed of fires. If combustibles are distributed vertically, the stratification of combustibles determines whether a crown fire or underground fire occurs.

③ Size of combustibles: The size of combustibles affects the flammability of such combustibles. Under the same heat source, the ignition time of combustibles with a larger size is longer than that of combustibles with a smaller size. However, the combustibles with a larger size can produce higher energy.

(2) Topographical conditions

① Slope: With the increase of slope, the spread speed of a forest fire spreading uphill increases. Generally, for every increase of 15° in slope, the spread speed of forest fire increases by one time. The spread speed of a fire on ridge is faster than a fire on slope. When the slope exceeds 40 degrees, the spreading of the uphill fire shows leap-forward development, and the spread speed increases considerably.

② Slope direction: If slope directions are different, the ignition time and spread speed are different. For example, when compared with shady slope, the sunny slope has stronger sunshine and faster evaporation, and most of combustibles are positive plants which have lower water content and can be dried quickly. Therefore, fires on sunny slope feature shorter ignition time, higher spread speed and higher intensity.

(3) Meteorological factors

① Precipitation: Precipitation can temporarily reduce the combustibility of forest combustibles, or cause such combustibles lose their combustibility.

② Relative humidity: The relative humidity can change the water content of forest combustibles. Therefore, changes in relative humidity can directly affect the combustibility of forest combustibles.

③ Temperature: When the temperature is high, the forest combustibles are dry and flammable, so the fire can spread very quickly in such case. Moreover, temperature affects relative humidity.

④ Wind: Wind can accelerate the water evaporation of forest combustibles and bring a large amount of oxygen which can support the combustion of forest combustibles. The wind determines the spreading direction and speed of a forest fire. When the wind speed reaches a certain level, the wind can change the direction and the height of the convective columns (smoke columns), and blow burning objects to other places, thus causing flying fires.

(4) Fire behavior: At present, the extinguishing of a forest fire, no matter whether such fire is in the southern or northern forest areas in China, is mainly dependent on the group fire brigades (including the extinguishing operation by helicopter). As a result, it is very necessary to take corresponding tactical countermeasures according to the forest fire behavior.

① Surface fire. With a slow spread speed, creeping surface fire features full burning of combustibles, regular fire field and clear fire line, and is easy to control. Rushing surface fire features rapid spread (the speed can be several times higher than 1 km/h) and uneven burning of combustibles. Unburned grounds are often left, resulting in repeated burning or the phenomenon that large fire field is surrounded by small fire scene. Therefore, the real fire line is hard to judge, thus bringing great difficulty to the fore fire suppression.

② Crown fire. The ground fire brigade cannot directly put out crown fire, but must adopt indirect extinguishing methods. In addition, high importance shall be paid to personal safety during the extinguishing process. The crown fire features rapid spread speed, high fire intensity, extremely irregular fire line, and changeable directions of fire spreading.

③ Underground fire. Underground fire is hidden and difficult to detect. Sometimes, even no smoke is seen. Compared to the extinguishing of surface fire, the extinguishing of underground fires not only adopts different tools, but also use different methods.

(5) Condition of fire field: The fire condition means the size, shape and location of the fire field. The size of the fire field decides how many firefighting forces shall be input. The

bigger the fire field or the more the fire brigades, the higher the attention shall be paid to cooperative work and tactics.

The shape of the fire filed determines the way how to call the fire brigades. The fire lines shall be regular to facilitate the extinguishing, cleaning and guarding operations of the fire brigades. If the fire lines are messy, the real external fire edge must be identified. Never extinguish the fire within the fire scene as that outside the fire scene, or otherwise, the fighting capacity will be wasted and safety issues will be caused.

The geographical location of the fire scene determines the application of firefighting tactics. It shall be well known that what influence will be imposed by the terrain on a fire and which place will accelerate, slow, strengthen, weaken or block a fire, or cause fire danger.

(6) Firefighting strength: Quantity and quality are two necessary elements for the firefighting strength of a fire brigade. Without certain quantity, quality is out of the question. Quantity is the basis of the quality. In return, quantity doesn't represent quality, and quality is the inherent potential of the fire brigades. Therefore, during the forest fire suppression, a huge-crowd strategy without limitation shall be avoided, but the labor force shall also not be in shortage. For forest fire suppression, the input labor force depends on the actual fire situation, and the quality of fire brigades will determine the quantity of fire brigades to be involved.

The extinguishing tools are the weapons of the firefighters. Different tools have different performances. The extinguishing machines and tools also determine the extinguishing modes and methods. For huge fires burning on the ground, water tender or helicopters designed to sprinkle chemical agents can be used to conduct direct extinguishing operations, while the use of No. 2 tools or pneumatic extinguishers means only the indirect extinguishing modes are available. Pneumatic extinguishers can easily extinguish medium/weak surface fire, but have no effect on the underground fire which can only be extinguished with a shovel or mining equipment.

Command and coordinate command means the determination and implementation of strategic and tactical countermeasures by the commander. Coordinate command include the leadership of the commander on the team and the initiative compliance of the fire brigades.

This reflects the comprehensive completeness of the fire brigades. The commander is the key to determine whether the fire brigades can give full play to their potentials.

3. Principles of determining tactical countermeasures

(1) "People oriented, safety first": The tactical countermeasures shall adhere to the principles of "people oriented, safety first". During the forest fire control work, the priority shall be given to the safety of lives and properties of people living in the forest areas and the fire crews and brigades working on the front lines of firefighting. Efforts shall be made to reduce or eliminate casualties caused by forest fires.

(2) Align subjective guidance with reality: Subjective guidance is tactical countermeasures that have been established. The tactical countermeasures must be based on the specific cleaning in four terms: Weather, land, fire and human beings. The situation of the fire scene is changeable. There are no such two fire scenes which are exactly the same. Different fire scenes shall adopt different modes and methods for fire suppression. In addition, the extinguishing tools used in the equipment of fire brigades called to different fire scenes are different. Different brigades also mean different leading methods. Therefore, we must know our opponents and ourselves and align the subjective guidance with the objective reality.

(3) Strengthen the ideological and political work and keep strong morale. Keeping strong morale is the fundamental guarantee for the success of forest fire control. Actual fire extinguishing practices have proved that the extinguishing environment is very tough, and sometimes even very cruel. In such case, in order to make the fire brigades united, fighting bravely, unyielding, unbeatable, unbreakable and invincible, efforts shall be made to keep the strong morale of the fire brigades. Strong morale comes from effective ideological and political work. Different from daily ideological and political work, the ideological and political work during the forest fire control process needs to be done in special modes and with special method under certain environments. Generally, the following requirements must be met:

① The commander shall explain situation, clarify tasks, and give clear requirements;
② The commander must practice what he/she preaches. For anything he/she asks others to

do, he/she must do first;

③ The commander shall act as both the leader and the member of the fire brigade;

④ The commander shall pay attention to the risks, and protect the personal safety of team member everywhere;

⑤ The award and the punishment should be proper, and public interests shall be separated from private interests.

(4) Give commands in a unified way and uphold strict discipline: During forest fire suppression, uniform command must be emphasized. The implementation of firefighting tactical countermeasure should be completed by the fire brigades. The commander himself shall lead all fire brigades as flexibly as he/she control his/her fingers. The specific operations shall include:

① The commander and fire brigades have to know and under each other;

② The orders and instructions must be consistent and clear, and frequent changes in policy shall be prohibited;

③ High attention should be paid to strike a proper balance between work and rest of fire brigades, preventing wasting their energy and stamina;

④ The commander should not only be good at exercising his/her power endowed by his/her position, but also consciously establish his image in practices, so as to win trust and respect from others;

⑤ Strict discipline and clear division of labor shall be ensured;

⑥ High importance should be attached to the communication.

(5) Concentrate superior forces to fight a war of annihilation: The superiority must first be established on preparedness. The so called "preparedness" includes the preparedness in terms of team, thinking, equipment and training. Only fighting a war of annihilation can ensure that there will be no re-ignition on the fire scene. In local sections, superior forces can be concentrated to put out a fire completely. Especially when the extinguishing force is insufficient, we shall pay more attention to the superiority in time and space, and solve specific problems thoroughly. The war of annihilation is based on superiority, and the superiority shall be reflected in the following aspects:

① The commander and the firefighters have a perfect mastery of the teamwork spirit and can work together in a productive way;

② The commander and the firefighters are robust, strong and energetic;

③ The commander and the firefighters are familiar with the laws of forest fires and the skills in extinguishing forest fires;

④ The fire brigades have been equipped with handy tools and equipment.

(6) Seize opportunity and fight a quick battle: Once happening, the fire expands rapidly with the time, and spreads faster and faster. Under the influences of wind direction, wind speed, terrain and other factors, the fire intensity is changeable. The longer the fire brigades stay on the fire scene, the more tiered they will become. The prolonging of the extinguishing time means the increasing loss of resources. Therefore, all opportunities shall be seized during the firefighting process to use and create all favorable conditions to extinguish the fire quickly.

Quick action means to gather the team quickly, arrive at the field quickly, know the fire situation quickly, formulate plans quickly, issue instructions quickly, and dispatch team quickly.

The opportunity refers to the time when the fire just occurs, or the fire is small, weak, against the wind or downhill, or the fire is surface fire in forests, passing fire, cross-rive fire, or a fire that spreads near to the defensive line, or when the fire scene has high humidity and low temperature, or when the fire occurs at night.

(7) Value efficiency and minimize the damage: The forest fire suppression shall pay attention to economic benefits, strive to minimize the input and achieve the best results, and avoid that the loss outweighs the gain.

(8) Have the overall situation in mind and grasp the key points: When formulating tactics, the commander must pay full attention to get the big picture of the entire fire scene and know the overall situation and key parts.

4. Tactics

(1) Basic extinguishing tactics: The basic tactic of extinguishing forest fire is "dividing the

forces to encircle".

"Dividing the forces to encircle" means to break one or more points on the fire scene, and divide the firefighters on each broken point into two parts which need to extinguish the fire along different directions of the fire lines. After extinguishing and cleaning embers, guarding personnel shall be left at each point until all fire brigades join together to encircle the entire fire scene and extinguish the fire completely.

Attention shall be paid to three key points when using the tactic of "dividing the forces to encircle":

① The fire line to be extinguished must be the real external fire line;
② Breaking points shall be selected correctly and the troops shall be assigned appropriately;
③ The encirclement shall be tight, without any breach.

(2) Common tactics

① Using the tactic of "dividing the forces to encircle" in combination with the deep-thrust battle: After "dividing the forces to encircle", dispatch part of fire brigades on the fire scene to the opposite or lateral fire lines, and require these fire brigades to break such fire lines and form a new encirclement. When this tactic is used, a fire brigade or more can be dispatched to fight the deep-thrust battle, according to the fire scene and the actual situation of the fire brigades. Attention shall be paid to the following problems when using this tactic:

a. Each fire brigade shall know the situation of the fire scene and have accurate judgment, and once this tactic is used, the major fire brigades must aim at the key part and concentrate a superior force to extinguishing the fire thoroughly in such key part; or
b. Each fire brigade shall be responsible and never give up;
c. Each fire brigade shall know the geographical situation and road conditions on the fire scene and shall never lose their way. Failure to reach the appointed place will spoil the opportunity.

② Using the tactic of "dividing the forces to encircle" on a line formed by two given points: When the closed form of original fire lines on the fire scene are broken due to terrain, forest combustibles, weather conditions, blocks and human firefighting behaviors and one or more fire lines appear, for one certain fire line, the fire brigade can conduct extinguishing work from the two end points of the fire line and gradually move to the middle point. When the two brigades join, the fire line is completely put off.

③ Push-in extinguishing: With this method, the fire brigades, after entering the fire scene, shall start the extinguishing operations from one end of the fire line straightly to the other end. This method can obtain very good result when the fire line is short and the fire is weak.

④ Progressive extinguishing: This tactic is similar to the "push-in extinguishing", so the fire brigades also need to start the extinguishing operations from one end of the fire line straightly to the other end. The difference is that: The order of the firefighters during the push-in extinguishing process hasn't changed, while the progressive extinguishing requires the firefighters to be divided into several teams and these teams will march forward alternately after they enter the fire scene.

This method is generally applicable to large fire scene or high fire intensity and a great number of firefighters. This tactic can increase the efficiency of the fire brigades and save the labor of the firefighters, because firefighters can get the change to take a short break when marching alternatively.

⑤ Guerrilla tactic of defeating one by one: This method is often used at night or early morning. As the fire observation is extremely clear at night and the night or early morning features low temperature, high humidity and small wind, some part of the fire line often extinguish itself. As a result, the enclosed fire line becomes intermittent and disconnected segments. Then, the fire brigades can be divided into several teams, and each team shall move according to the general direction to complete their tasks. Therefore, these teams extinguish the fire during the movement.

Using this tactic, the following points shall be bore in mind:
a. Each team must be active and initiative;
b. Each firefighter must be an individual who has received relevant training, had rich experience and can conduct the extinguishing operation independently;

c. The fire brigade must have good communication equipment and can communicate effectively on the fire scene.

Ⅴ. Extinguishing methods

The extinguishing methods can be divided into two categories: A. Direct extinguishing methods, such as extinguishing by beating and flapping, extinguishing with soil, extinguishing with water, extinguishing with wind and extinguishing with chemical agents; B. Indirect extinguishing methods, such as digging fire trenches, setting fire breaks, extinguishing with fire, and extinguishing by explosion.

1. Extinguishing by flapping

Flapping is the most primitive and common method to extinguish a fire. This method, i.e., use the handheld extinguishing tools (such as No. 1 tool and No. 2 tool) to flap the fire directly, is applicable to extinguishing weak and medium surface fire.

Using this method, firefighters shall stand at a certain angle against the fire line and align No. 2 tools oblique to the flame so that the angle between the tool and the ground can be 45 degrees. Slight lifting, forceful pressing, flapping and dragging can put out a fire easily. An angle of 90 degrees between No. 2 tools and the ground, straight flapping, violent actions shall be avoided, or otherwise these actions will fan the combustion and cause splashes, forming new ignition points.

2. Extinguishing with soil

This method is applicable to forests with thick litter layer and many messy things. When the flapping method is not applicable, a hoe, a spade and other tools can be used to take earth to cover the fire. This method generally is used when the forest soil is loosened. This method has the advantages of reachable materials and good effect. When soil is used to extinguishing embers during the clearing stage, this method is very effective to prevent resurgence.

If the fire spreads faster than soil covering rate, an open fuel break shall be set up ahead of the fire heading to stop the spread of the fire. The width of the open fuel break shall be based on the vegetation types and the coverage rate. For example, when the litter layer is thin and many moss grow in the forests, the width of the open fuel break is generally 0.5–1.5 m. When the litter layer is thick in the forest or there are many fallen trees, the width of the open fuel break shall be 2 m or above. If there are thick gramineous weeds on the ground of the forests, the width of the open fuel break shall be 4–5 m in order to stop the spread of the fire.

3. Extinguishing with water

Water is one of the most commonly used extinguishing agents. In nature, there are abundant water resources. If there is a water source near the fire scene, the fire shall be extinguished directly with water. In this case, this method can not only shorten the extinguishing time, but also prevents resurgence.

The principle of extinguishing with water: First of all, water cools the temperature. Water has a great heat capacity, and 1 kilogram of water needs to absorb 4.19 kJ of heat for every increase of 1 centigrade degree. When water evaporates, it needs to absorb 2,558.41 kJ of heat for vaporization. When water absorbs a large amount of heat from the burning material, the humidity and flammability of such material increase. Second, the water vaporizes after being heated, 1 liter of water can become 1,500 to 1,720 liters of water vapor which can dilute the concentration of oxygen in the air and reduce the amount of oxygen added so as to achieve the purpose of extinguishing the fire. Third, the water column ejected under pressure has a certain mechanical function, and can destroy the burning dead branches and leaves, and mix them with soil, thus extinguishing the fire.

Machines used for extinguishing with water mainly include: Water gun extinguishers, backpack fire extinguishers, mobile relay pumps, forest fire vehicle and sprinkling airplanes.

4. Extinguishing with wind

Extinguishing with wind refers to use strong winds provided by pneumatic extinguisher to blow away the heat released by the combustion, reduce the temperature to below the ignition point, and to dilute the combustible gas to a concentration lower than the combustible concentration, thus putting out the fire. The pneumatic extinguisher generally can put out weak and medium surface fire, rather than hidden fore or crown fire. The pneumatic extinguisher can be mainly used in the following ways:

(1) Flapping: Pneumatic extinguisher is mainly used to flap out weak, medium, and strong surface fire, and serving as one major method of cutting flame button lines.

(2) Cleaning: Pneumatic extinguisher is mainly used for cleaning the fire line and widening fire breaks.

(3) Controlling: Pneumatic extinguisher is mainly used to suppress flames and prevent the expansion of the fire line and combustion.

(4) Cooling: Pneumatic extinguisher is mainly used to eliminate the threat of the heat radiation on the fire extinguisher and fan fuel tanks. When in use, the pneumatic extinguisher can be used independently, in pairs or in triple in combination with water guns and other extinguishers.

5. Extinguishing with fire

It is an effective method for extinguishing high intensity surface fire or crown fire. When the fire spread fiercely and firefighters cannot directly extinguish the fire, set fire breaks ahead of the fire head, and then, ignite a fire toward the spreading direction of the fire head. Due to the fire breaks, the ignited back burn fire spread only to the fire head. Therefore, when the two fire heads meet, the fire will extinguish itself. This method requires no special equipment, and if handled properly, can save time and labor and increase extinguishing efficiency. However, this method has high operation requirement and is hard to master. If used improperly, it will not extinguish the fire, but increase the fire intensity, accelerate the spreading of the fire, expand the fire scene, damage the whole extinguishing plan, and even cause casualties. Therefore, commanders with rich practical experience must be required to command. This method is an emergency method used in circumstances where the fire is strong, the natural barrier and artificial fire facilities are unable to prevent the spread of fire and there is no time to set wide fire breaks. This method is generally used in the following two ways:

(1) Burning method: When the fire occurs, the air in the fire scene is heated and rises, so that the surrounding air moves to the fire direction, producing cyclones. The ignition outside the cyclone formation zone but ahead of the fire head is called the burning method. This method uses the fire breaks, natural barriers and roads as the control line, and ignites to the fire head so that the fire will burn to the fire scene against the wind. When the two fire heads meet, the fire will extinguish itself. It can be divided as single-layer burning method and multi-layer burning method. If a single layer of burning cannot stop the fire head, another burning layer can be used, and the fire breaks are widened to stop the spread of the fire. This is called the multi-layer burning method.

(2) Back burn method: The back burn method is to ignite within the cyclone formation zone but ahead of the fire head so that the fire spread to the fire head and extinguish itself when the two fire heads meet. The determination of a cyclone can be determined by observing the direction of the smoke or throwing paper slips or watching the flying direction of leaves. The back burn method uses the fire breaks, rivers and roads as the control line, and ignites lither ahead of the fire head when a cyclone forms in the fire scene. If the ignition happens before the formation of the cyclone, an opposite results may be obtained, or a great danger may be caused to fire fighting personnel. Therefore, special caution should. be taken to select officers who have sufficient fire fighting experience to give command on the spot.

6. Extinguishing by explosion

The explosion method can not only open fuel break and fire trench, block the spread of fire, but also use the shock wave produced by the explosion and soil to put out a raging fire directly. Generally, this method can be used when a large area of fire occurs in remote forest areas, and firefighters are inadequate, or forest clutter is heavy, or forest soil is hard. There are some explosive methods, such as hole explosion, explosive blasting, dry powder extinguishing bomb, extinguishing gun and so on.

7. Retarding with fire breaks

The terrain in the southern area is complex. The forest fire behavior is changeable due to mountain terrain and weather conditions (wind direction, wind strength).

Fierce uphill fires are difficult to extinguish by flapping. In this case, fire breaks may be set in a favorable terrain or vital position with shovels, hoes, hatchet, axe, saw and other hand tools at a place at a certain distance from the fire head or by the sides of the fire wings. Weeds and litter shall be removed, and flammable woods shall be cut and moved to a safe place. If time allows, a temperature fire break, open fuel break, or fire ditches shall be evacuated to stop the spread of the fire and extinguish the fire by isolating it.

The fire breaks shall not be set in the uphill direction of the uphill fire, because the uphill fire spread quickly and are very dangerous. Most of fire breaks are set ahead of the downhill fire. If the uphill fire spread slowly, and the terrain is gentle, fire breaks can be set along the ridge line, blocking forest fire into a downhill fire.

8. Extinguishing with chemical agents

Extinguishing with chemical agents is a method of extinguishing fire or blocking fire spread by using chemical agents or powder. This method has the advantages of fast extinguishing rate, good effect and low recurrence rate. Everywhere the chemical agents arrive, the fire goes out immediately. This method has changed the traditional method of extinguishing fire with the crowd tactic, saves labor, and realizes the goal of "extinguish a fire in early stage, extinguish a fire when it is small, and extinguish a fire till it completely goes out". The method can be used to directly extinguish forest fires such as surface fire, crown fire and underground fire, and can also be used for setting fire break. Especially in the remote and inaccessible forest areas, it is a modern and advanced measure to use aircraft to spray chemical agents to directly extinguish or block fires.

(1) Principle of extinguishing chemical agents

① Moisture absorption mechanism: Some extinguishing chemical agent has high moisture absorption performance. Even at high temperatures, it also has the ability to keep water. Keeping the water content of the combustibles can reduce the combustibility of the combustible. For example, ammonium chloride loses its crystalline water at 250℃, while wood begins to burn at 230℃ to 300℃, so ammonium chloride can be used to extinguish fire.

② Dilution mechanism: Some extinguishing chemical agents can exude difficult-combustible gas (such as steam, nitrogen gas, carbon dioxide, and ammonia gas) when heated, these gases can dilute the concentration of combustible gas and meanwhile can reduce the concentration of oxygen in combustion regions, thus extinguishing fires.

③ Covering mechanism (coating theory): Some extinguishing chemical agents can form a layer of film when heated and decomposed, and the film will cover the surface of burning items which will insulate oxygen supply, and further restrict combustion process from spreading, thus extinguishing fires finally.

④ Heat absorption mechanism: Some extinguishing chemical agents will absorb a great amount of heat energy when dissolved, which will reduce the temperature of combustion regions, weakening combustion process.

To sum up, extinguishing chemical agents shall at least have one of the following characteristics: High hygroscopicity; release combustion-inhibit gas when decomposed; form films when heater; absorb a great amount of heat energy when dissolved.

In the selection of extinguishing chemical agents, you shall not only consider their effects in fire extinguishing, but also shall consider if such agents have abundant resources and low prices, and if they are reliable, non-toxic and non-corrosive etc.

(2) Extinguishing chemical agent

① Types of extinguishing chemical agents: The agents consist of long-acting fire extinguishing agents and short-acting fire extinguishing agents. The former regard water as the carrier which rely on chemical agents to extinguish fire, and these chemical agents still have the function of inhibiting combustion after the water has completely evaporated. Such chemical agents, can firmly adhere to combustible materials after being applied, are not afraid of being washed off by rainfall and can maintain effective fire prevention and extinguishing within a few months. Long-acting fire extinguishing agents mainly include various types of fire extinguishing agents which take ammonium phosphates or ammonium sulfates as main agents, such as Fhos-Check, Fire-Trol, "704" and "75" fire extinguishing agents.

Short-acting fire extinguishing agents mainly rely on changing the nature of water (dilute and thicken) by chemical substances to extinguish fire which cannot extinguish fire and prevent fire once lost water. Such agents, cannot resist being washed off by rainfall after being applied, cannot adhere to upright combustible materials for a long term, therefore they cannot maintain long-term effective effect. Normal thickening agents include: Bentonite, algae gum, the master batch of Gill Celgard Jet and carboxymethylcellulose sodium etc.

② Dosage form of chemical agents: Liquid fire extinguishing agents, such as water; turbid liquid fire extinguishing agents, such as “704” and “75” fire extinguishing agents and Fhos-Check; emulsion fire extinguishing agents, such as Freon; Foam fire extinguishing agents, such as fluoroprotein foam fire extinguishing agents; Gas fire extinguishing agents, such as carbon dioxide, smoke and fog fire extinguishing agents; dry powder fire extinguishing agents, such as: Sylvite dry powder, sodium salt dry powder; block-type fire extinguishing agents, such as: Explosive.

③ Main ingredients of extinguishing chemical agents: An extinguishing chemical agent generally contains the following ingredients:

a. Main agent. It is an agent that has an effect of extinguishing and preventing fires, such as ammonium sulfates and ammonium phosphates.

b. Additive agent. It is also called reinforcing agent, which has an effect of enhance and improve the fire-extinguish effects of main agents. The cooperation of additive agents and main agents can improve the fire-extinguish effects of fire extinguishing agent. Such as add a certain quantity of ammonium bromide in ammonium phosphates can improve the fire-extinguish effects of ammonium phosphates.

c. Wetting agent. It is an agent which can not only reduce the surface tension of water, but also improve infiltrating and spreading capacity of water and cause emulsification and foam effects at the same time.

d. Viscosity agent. It is a kind of chemical agent that can enhance the viscosity and adhesion of fire extinguishing agents, and reduce the wastage and dissipation of fire extinguishing agents. Common viscosity agents include: Pectin, bentonite and activated carclazyte etc.

e. Anticorrosive agent. It is an agent that can prevent and reduce the ingredients of fire extinguishing agents that can corrode metals. Such as potassium dichromate.

f. Antiseptic. It mainly prevents bacterial contamination to reduce the viscosity of extinguishing chemical agents significantly. Therefore it shall add a little of formaldehyde in fire-prevent chemical agents as antiseptics.

g. Coloring agent. Fire extinguishing agents are always added with dyes or pigments to identify the regions that have been sprinkled with fire extinguishing agents. The color of dyes or pigments is normally red or blue, such as acid red and iron oxide etc.

(3) Common extinguishing chemical agents

① Ammonium phosphate-type extinguishing agent: Including ammonium dihydrogen phosphate (MPP), diammonium hydrogen phosphate (DAP), phosphate fertilizer, ammonium pyrophosphate and ammonium polyphosphate (APP) etc.

"704" extinguishing agent: An extinguishing agent developed in 1970s in China, which is consist of the following ingredients: 29% of phosphate fertilizer, 4% of carbamide, 1.3% of sodium silicate, 2% of washing powder, 0.25% of potassium dichromate, 0.1% of brilliant crocein, and 63.35% of water. This extinguishing agent is read turbid liquid with pH value being 6.8, density being 1.22, viscosity being 0.01–0.05 Pa.s (10℃) and surface tension being 0.0483 N/m (10℃).

② Ammonium sulfate-type extinguishing agent: White or slight yellow crystals, highly water-soluble, insoluble in alcohols.

③ Extinguishing agent dominated by ammonium sulfates and ammonium phosphates: Recent researches indicate that, the extinguishing agent with main agents being ammonium sulfates and ammonium phosphates has synergy effects to some extent, thus improving the fire-fighting efficiency of fire extinguishing agent, reducing the cost of the agent and its corrosive action to metals.

"75" extinguishing agent made in China: With main agents being ammonium sulfates and ammonium dihydrogen phosphate, and is consist of the following ingredients (% weight of ammonium): 28% of ammonium sulfate, 9.3% of ammonium phosphate fertilizer, 4.7% of bentonite clay, 0.9% of trisodium phosphate, 0.9% of washing powder, 0.1% of brilliant crocein and 56.1% of water.

Besides, there are also Fhos-Check-D_{75}, Fire-Trol-GTS and Fire-Trol-PSF.

④ Halide-type extinguishing agent: It can be divided into inorganic halides and organic halides. Common inorganic halides include: $CaCl_2$, NH_4Cl, $MgCl_2$, $ZnCl_2$ etc. Common organic halides include: Freon and syngnathus. Freon has not been used in forest fire control due to its high price.

⑤ Foam fire extinguishing agent: An agent made by adding foaming agent in liquid fire extinguishing agent, such as prion protein and compound foaming agent. The foam can adhere to combustible materials to form pellicular insulating layers which can insulate air and energy, thus suffocating fires.

⑥ Smoke extinguishing agent: Smoke refers to solid molecular groups or granulum that disperse in the air, fog refers to liquid water droplets that disperse in the air. Main ingredients of this agent include: a. Fire extinguishing agent: 35%–60% of halides; b. Exothermic agents: 20%–35% of oxidant, such as potassium chlorate, sodium perchlorate, potassium perchlorate and ammonium perchlorate etc.; c. Adhesive: 30%–35% of epoxide resin, polyurethane etc.; d. Weighting material: 4%–33% of sheet sand, ferrous powder and titanium dioxide etc.

⑦ Dry extinguishing powder: A thermosynthesis product of sylvite and urea, such as China's sodium salt dry powder.

⑧ Explosive-block fire extinguishing agent: An agent which utilizes the numerous nonflammable gas and intensive blast wave which are produced at the moment of the detonating of ammonium nitrate, TNT and black gunpowder, to insulate air, dilute the percentage of oxygen and eliminate main combustible materials.

(4) Use of extinguishing chemical agents

① Sprinkle extinguishing chemical agents from airplanes. This method can be divided into two methods: Sprinkle through dumping to put out fires directly and sprinkle fire control strips to prevent fires from spreading.

② Sprinkle through ground-based machines and tools, i.e. sprinkle with forest fire truck, fire extinguisher etc.

③ Extinguish fire with explosive, i.e. extinguish through explosion.

(5) Precautions on the use of extinguishing chemical agents.

①Extinguishing chemical agents are generally toxic. If they are used in a large amount, then they will cause certain effects on the biological and environmental factors in forest ecosystem, and bring certain pollution to air and water source, therefore negative effects after the use of such fire extinguishing agents must be researched.

②These agents are toxic, so if they splash on the skin, they may irritate the skin and may result in a feeling of "burning", therefore users must pay attention to safety.

9. Extinguishing with aerial machines and tools

In remote forest region with inconvenient transportation and few inhabitants, utilizing airplanes to extinguish forest fires is fast, high efficient and flexible, which has an important strategic meaning. Aeronautical fire-fighting methods include: Extinguish fire through a parachute, extinguish by aircraft landing, extinguish fire through fast roping, sprinkle water from air and extinguish fire through sprinkling chemical agents.

(1) Extinguish by parachuting: Utilize fix-wing aircraft to extinguish fire through a parachute near fire scene. The biggest advantage of this method is that: Firefighters can observe fire conditions promptly, thus putting out newly-rise small fires immediately, and then preventing small fires from spreading. Major tasks of airborne personnel are:

① Extinguish fire directly: Organize ground crew to extinguish fire.

② Parachuting leads the way: In the case that firefighters have approach the fire field but cannot find such fire field or firefighters have lost directions after a firefighting, airborne personnel may lead firefighters to recognize right directions and reach destinations immediately through a parachute.

Extinguishing by parachuting is a relevant complex and hard work, and the complex terrain of forest regions and changeable weathers may bring difficulties to the landing of a parachute. If the landing ground is not selected properly, then the parachuting is prone to make personnel injured, thus affecting the work of firefighting directly.

(2) Extinguishing by transporting firefighters with helicopter: Utilize a helicopter to transport ground firefighters immediately to fire scene to extinguish fire. The advantages of such method are:

① Can transport firefighters immediately to fire scene, which can not only reduce the physical consumptions of firefighters due to long journey, but also can maintain fire fighter's combat power;
② Convenient to mobilize and adjust fire-fighting forces, as well as move to other fire scenes and intercept main fire heads;
③ Safe, accurate and efficient in firefighting;
④ Less investments are required for venue construction.

China has started to utilize helicopters to extinguish fire since 1965. It mainly consists of specialized fire-fighting teams and supplemented by local organizations, aircraft-landing fire-fighting teams regard extinguishing initial small fire scene as main subjects, and the teams are equipped with the most advanced fire-fighting machines and tools, and make various fire-fighting methods such as patrols, a watch from high places, aircraft landing, ground transportation and aeronautical chemical firefighting coordinate closely.

(3) Extinguishing by landing firefighters with ropes from helicopters: A method to help firefighters to land from a hovering helicopter to extinguish fire directly or open up aircraft landing ground in shot time through auxiliary equipments such as winch devices, steel

rope and harness system, thus helping large quantities of aircraft-landing team members to touch down. This method can make up for the requirements on landing sites by aircraft landing, and firefighters can extinguish fires directly in a place that has no ground for aircraft landing.

(4) Extinguishing with hanging water buckets: It is a new type of fire-fighting methods that extinguishes fires through sprinkling water or chemical agents with a bucket that is hung on the outside of a helicopter. This method has been widely adopted in some developed countries like Canada and Japan and the forest region of southwest China and northeast China. Such bucket has metal frame and canvas enclosure, its upper part is opened for water (or chemical agents) and its bottom has a valve which is used to release controlling devices to connect with releasing valves and the control cabin. The shape of the bucket is varied, and in Japan and northeast China, it is always conic with a carrying capacity of around 1,000 kg. The bucket is removable and collapsible, and is easy to transport and carry.

Using a bucket on a helicopter to extinguish fire can be divided into two ways: Direct firefighting and indirect firefighting. The former is the common method that sprinkles water (or prepared fire-fighting solutions) in buckets directly on fires. Such buckets can sprinkle water into band shape, strip shape or block shape in accordance with fire intensity and technical requirements of water sprinkling operations. This method can be used to control such fire heads that spread faster or such fire heads that has strong combustion intensity and is nearer to firefighters. The latter is to release water in water tanks on the ground firstly, and then firefighters draw waters from the water tank to fire lines to extinguish fire, or prepare fire-fighting solutions in proportion in water tanks firstly, and then draw prepared solution to fire lines to extinguish fire.

Transfer water (or prepared fire-fighting solutions) through water pumps and fire hose to fire lines to extinguish fire, when compared with extinguish fire through sprinkling waters from buckets, has a higher water utilization and better effects. Using a bucket on a helicopter to extinguish fire is more accurate and has higher water utilization than the method of extinguishing fire with fix-wing aircrafts, and the former method has low requirement on the conditions of water source which the seas, reservoirs, rivers and ponds can also be the water source for firefighting.

(5) Extinguishing by sprinkling from airplanes: A method that uses airplanes to convey waters or chemical solutions to extinguish fire by sprinkling such water or solutions above fire scene. If this method is adopted, a water pond must be built in the districts where water sources are insufficient, or a chemical fire station can be set at the landing sites of airplanes, or a temporary station for the preparation of chemical agents can be set near fire scene, as supplements of fire-fighting solutions in airplanes.

10. Extinguishing fire through artificial catalytic precipitation

Artificial catalytic precipitation is not only an effective fire-fighting method but also a preferable fire prevention method. Pre-commit an artificial catalytic precipitation in forest regions that is arid and has higher fire hazard potential prior to a fire hazard can effectively reduce the combustibility of such forests, thus preventing a forest fire. Initial an early precipitation or increase the intensity of precipitation in areas that is experiencing a fire hazard can extinguish forest fire or reduce fire intensity which is favorable for ground fire-fighting.

During the period that a fire hazard is potential to happen or has happened, a rainy weather always appear, but the rain will not fall since it has not reached the critical point. Summer precipitation generally generate ice crystals above supercooling cloud layers, the ice crystals will enlarge while absorbing moisture, and they will fall into warm airs and form rains when airflows cannot burden them. Artificial catalytic precipitation is to add catalysts in cloud layers to promote the functions of ice crystals, thus promoting a precipitation.

Catalysts for an artificial precipitation mainly include dry ice, silver iodides, aluminum iodides and cupric sulfides etc. The metaldehyde that is researched and manufactured by University of Denver in U.S. has a price that is 1/100 of silver iodides, but its capacity of generating ice crystals is 1,000 times higher than silver iodides.

There are 4 methods to sprinkle catalyst to cloud layers:

(1) Crush dry ice into blocks that are less than 3 cm, then sprinkle such blocks in the cool cloud layer from airplanes, and the blocks will evaporate while descending, thus generating ice crystals.

(2) Put a chunk of dry ice into metal-filament containers, and then fix them on the surface of a fuselage, and such dry ice will evaporate into clouds when the airplane is flying, thus generating ice crystals.

(3) Launch through an antiaircraft gun or a rocket.

(4) Bring dry ice into clouds with balloons.

Artificially promoted precipitation can be realized only through a certain of meteorological conditions, i.e. a certain of cloud layers. Some falling woods or remains of logging in some specific areas of fire scene may still burn after an artificial precipitation, therefore it is necessary to dispatch firefighters to extinguish fires and clear fire scene even after an artificial precipitation.

In "5·6" extraordinary serious forest fire in the Greater Khingan Mountains in 1987, totally 10 artificial catalytic precipitation by airplanes were conducted, which had great effects on forest firefighting, causing attention from those in and out of China.

Ⅵ. Safety of firefighting

1. Safety rules for firefighting

(1) No disabled personnel, pregnant woman, minor or other personnel who are not applicable to participate in such activities shall be advocated to put out a forest fire.

(2) Firefighters must accept safety training for firefighting.

(3) Comply with fire field disciplines, uniform instruction and deployment. No solo action is permitted.

(4) Ensure your communication always being unobstructed.

(5) Firefighters are required to be equipped with essential equipment, such as helmet, fire-

fighting suit, fire resistant gloves and boots and fire-fighting machines and tools.

(6) Pay close attention to weather conditions of fire scene, especially during the early afternoon, a casualty accident-prone period when putting out forest fires.

(7) Pay close attention to the types and flammability degree of the combustible materials on the fire field, and avoid accessing to areas with flammable materials.

(8) Pay attention to the terrain of fire scene. Firefighters shall not enter into such places as follows while putting out fire head: A place which is surrounded on three sides by mountains, saddle valley, narrow grassy pond and ditch, narrow valley, sunward mountainside.

(9) Fire refuges and evacuation routes shall be selected prior to a forest-firefighting to against any incidents. Once trapped in dangerous environment, you should keep your mind clear and endeavor to save yourself.

While putting out an underground fire, you will find out the scope of the fire field and make some marks to avoid entering into such fire field.

(10) Considering firefighters' physical strength consumes greatly, therefore firefighters shall take appropriate rest to maintain vigorous physical strength.

2. Responsibility of the fire-fighting instructor

(1) Grasp timely the information of weather conditions of fire scene.
(2) Make accurate prediction and judgment on the development of the behavior of forest fires.
(3) Make adequate emergency preparation for various potential conditions that may incur in fire scene.
(4) Prearrange evacuation routes.
(5) Pay close attention to potential dangerous sections.
(6) Ensure all lines of communication always being unobstructed.
(7) Grasp timely the actions of fire-fighting teams and fire-fighting progress.
As an instructor of firefighting, you shall constantly remind firefighters the following issues while putting out forest fires.

① Fire refuges must be established;
② Direct fire extinguishing method can only be adopted in possible circumstances;
③ Do not beat the fire head;
④ Never leave things to luck in any circumstances;
⑤ Grasp timely the information about weather conditions of fire scene.

3. Causes of casualty accidents

(1) Evacuate with downwind.
(2) Evacuate to the mountain which has not been burned.
(3) Evacuate to saddle regions.
(4) Approach to downhill fire scene from the mountain.
(5) Climb over topographical crest or saddle regions to approach to fire scene.
(6) Approach to fire scene against wind.
(7) Misjudge the change of the behavior of forest fires.
(8) Evade fires in marshy grassland or miscellaneous shrub.
(9) Firefighters are overfatigue, or smoked or choked by heavy smoke, or burned with high temperature.
(10) Lower guard to small scaled fires which mislead people to weaken their awareness to safety. Casualty accidents are prone to happen if no fire refuge is built.
(11) Be extremely frightened and panic-stricken to forest fires. Firefighters without having accepted fire-fighting training may be too frightened and panic stricken to the booming noise, heavy smoke and the danger of high temperature arise from forest fires, thus causing casualty accidents due to a loss of uniform instruction or uncontrolled running.
(12) Injured by the falling woods burned by fires or random stones.

4. How to prevent casualty accidents

(1) Strengthen education work for firefighters.

① Strengthen the organization and leadership, and the members of fire-fighting team shall be those who are robust and strong other than those who are old, weak, sick, disabled, or the pregnant, or minors. Firefighters shall conform to disciplines, follow instructions and never leave the team without permission.

② Strengthen technical training and education on safety. Firefighters shall receive technical training and education on firefighting safety regularly. Stick to access to fire scene from fire rear, and follow the fire line to put out fires in the wing of fires, until the fire head is put out. Firefighters shall be brave but not be reckless and bravado during fire-fighting activities.

(2) Basic requirement for firefighters: Observe the weather conditions of fire scene, especially during the early afternoon, a casualty accident-prone period; pay attention to the change of terrain constantly, especially the change of slop aspect, gradient and position; comply with fire field disciplines, orders and shall not commit solo action.

(3) Safety during firefighting: Pay attention to the change of terrain while approach fire scene, keep away from narrow valley, steep slope, saddle regions, and grassy pond with tall and large grasses, and take care of the change of wind directions and speed.

Where isolation belts need to be set in front of surface fire and crown fire, safety refuges must be set. Burned areas are prohibited to access while putting out underground fires.

While putting out fire heads, firefighters shall have access to fire lines from the two wings of fire heads, and shall not put out the fire heads directly in front of such fire heads.

If firefighters are attacked by a blaze while putting out fire lines, they shall enter into burned areas immediately.

The following environment should not be selected as a campsite while camping: A place which is surrounded on three sides by mountains, saddle regions in a mountain ridge, narrow grassy pond and ditch, a steep slope at the foot of a mountain which has no plants and has loose sand or earth, or bare rock, a low-lying and dried riverbed and a place under an isolated big tree.

Where an emergent avoidance of danger needs to be adopted when the personal safety is threatened by big fires, firefighters shall firstly utilize favorable terrains, such as a highway, a railway, a path through the woods, a river and a stream, and then avoid dangers through making a fire at downwind regions; if there's no favorable terrain for avoidance, firefighters can make a fire at following wind regions and then enter into the newly burned areas to avoid dangers; if in an area where has no or rare plants and is near a river, a lake or a swamp, firefighters can utilize the terrain of such place to avoid danger. In the case that there's no condition to make a fire or avoid danger in other ways, firefighters may select a place with relevant plat terrain and lower fire temperature, and protect the head with clothes, then rush through fire lines against wind to enter into a burned area to avoid dangers.

5. Attention should be paid to the following major aspects when fighting forest fires.

(1) Approach the scene from the back and wings instead of facing it head on. During fire fighting, approach from two wings of the flaming front.

(2) Avoid approaching the scene from the top to the bottom of a mountain or by crossing a ridge or a saddle.

(3) Never set barrier zones above the scene or at ridge line when the fire is spreading rapidly up the mountain.

(4) When setting barrier zones, define or set safety zones and make visible the evacuation routes.

(5) After setting barrier zones, burn inflammables in front of fires at the edge inside the barrier zones in a planned and well-organized manner to widen the barrier zones.

(6) If there are rivers, streams, roads, paths or railways near the scene, burn down inflammables based on them under uniform organization; but take care to prevent fire spreading via bridges and culverts of such roads and railways.

(7) During the break, stay at the edge of the attack line where the fire has been put out.

6. Analysis on fire danger conditions

Firefighters should be familiar with severe weather conditions, disadvantage in terrain and dangerous combustible materials, understand forest fire behavior and changes and improve the ability to self-rescue in emergency.

(1) Severe weather conditions: Weather conditions leading to casualties mainly include increased wind force and sudden changes in wind directions. When the wind force reaches strong breeze, the fire will grow intensified and unable to be controlled by flapping; while sudden change in wind direction will easily result in casualties. At 12:00–14:00 when the air temperature is the highest and the humidity lowest, the combustion intensity is high and the fire spreads quickly. Fire fighting at such time may result in personal injury or even death; the improper route may also result in casualties.

(2) Adverse terrain: Adverse terrain refers to steep slopes, ridges, valleys with only one entry, ridge saddles and grass ditches. Such terrains may easily lead to sudden changes to the fire behavior and render it uncontrollable, thus resulting in casualties of firefighters. Since such terrains are located on sunny or half-sunny slopes, they are subject to long exposure to the sun, have high air temperature at daytime and combustible materials there are dry. Besides, some of such terrains may be low lying with low wind speed, therefore, fire head may easily turn. What's more, when the updraft interacts with the prevailing wind direction, strong convection may be produced, leading to fire whirl which can accelerate forest fire spreading and increase combustion intensity. If slope is not steep at such places, it may not be very for firefighters to advance, which may easily lead to misjudgment by the fire commanders and reckless actions of firefighters. When the fire crosses the ridge from the shady slope and spreads along the eastern or western slope down the mountain, the spreading is slow. Inexperienced fire commanders may mistakenly take such circumstance as the best opportunity for fire control and lead the team to the bottom improperly to attack the fire head. When the fire head spreads to the lower and middle part, the fire intensifies and turbulent fire and fire whirl start. At such time, the fire is fierce with large amount of smokes to darken the valley and flying sparks to ignite anywhere they go. Therefore, several fire lines and fire heads are created, downhill fire becomes intensive uphill spreading fire. Firefighters at the bottom may have nowhere to run and casualties may be

unavoidable.

(3) Combustible and inflammable materials: Combustible and inflammable materials are also the key to dangerous fire fighting conditions. The vegetation on sunny waste slope is over 1 m; the forest is dotted with inflammable bushes and oil-containing weeds that burn intensively and are not easily controllable; fire burns at the ground surface and tree crown inside a young coniferous forest; there are many combustible materials on a steep slope and it may easily result in vertical combustion, resulting in very dangerous fire conditions and casualties of firefighters. During fire fighting, special attention should be paid to distribution of combustible materials in trapezoid, heavy-loaded combustible materials, herbaceous combustible materials, inflammable bushes and young coniferous forest.

(4) Hazardous conditions: Camping outside scenes of fire; staying near scenes of fire; unable to see fire when fire existence is sure; unknown of fire location due to heavy smoke; sparks fly over to produce new fire points; unsure of fire condition even at a close distance with the fire; entering into cols of mountains; fire fighting near steep slopes and cliffs; air temperature rises; wind force intensifies; wind direction changes frequently; using indirect fire fighting methods without any retreat or getting lost.

7. Self-rescue in scenes of fire

Once firefighters are trapped in fire or attacked by spreading fire, quick decisions should be made to choose potential routes for breaking through or take proper fire avoiding measures to prevent accidents.

(1) Prevention and treatment of carbon monoxide poisoning: Firefighters have to work frequently in heavy smoke and therefore must learn to breathe in thick smoke and avoid carbon monoxide poisoning. When heavy smoke moves in your way, try your best to get out of the way. If it fails, get down to the ground with your mouth sticking to the ground (as low as you can) and breath. When exposed to dense smoke, try to move between smoke clouds and seek places with fresher air and lie down to breath.

(2) Treatment of trauma bleeding: In case of trauma bleeding, use fingers to press the wound to stop bleeding or use bandage or its substitute in the first-aid kit to dress; if the

injured bleeds heavily, use the tourniquet to stop bleeding and send them to the hospital as soon as possible.

(3) Treatment of fracture: The pain from fracture is usually unbearable. Hematoma or unmovable fracture places can be treated as follows: Bleeding wounds should be stopped bleeding; use splints to secure the injured places and if there are no splints, use wooden sticks or tree bark instead; do not bind the injured places too tightly; send the injured to the hospital immediately.

(4) Treatment of burns: Do not break the skin or blisters at burns and timely apply burn dressing; in absence of medicines, rinse the burns with water; if the burns are severe, send the injured to the hospital in a timely manner.

(5) Treatment of venomous snake bite: Remove venom at the bite as soon as possible and take antidote and sedatives immediately; if the artery is hurt and the bleeding cannot be stopped, use tourniquet and send the injured to the hospital quickly.

(6) Treatment of heatstroke at scenes of fire: Heatstroke at the fire scene happens quickly and its symptoms include extreme feebleness, dizziness, nausea, pale complexion, diaphoresis, skin clamminess, tachypnea and muscle clamminess or coma when serious. Treatment method: Move the patients to the shade and lay them on the back to rest; give saline water and sweet water to the patients; give fluid infusion if the patients lose too much water; send severe patients to the hospital.

(7) Self-rescue by burning down combustible and inflammable materials: If time permits, use rivers, streams or roads as retreats to rescue yourself by burning down combustible and inflammable materials in the way of fire spreading. If there are no rivers, streams or roads nearby, use igniters to create head fire and avoid fire inside burned area.

(8) Crossing fire line against the wind: If time is limited where advanced burning down or other fire avoiding measures cannot be used, remember never to escape along the prevailing wind direction. Choose locations with flat terrain, low fire intensity, thin fire wall and low flames, cover your head with clothes and quickly cross the fire line against the wind to enter burned area.

(9) Avoiding smoke (fire) by lying down: If time is not allowed to burn down in advance, but there are rivers, streams or places with no or a little vegetation, soak clothes with water, cover it with the head, place both hands before chest and lie down to avoid smoke (fire). When lying down, use wet towels to cover the mouth and the nose to prevent choking to faint or suffocation by the smoke.

(10) Quick evacuation: If big fire breaks out menacingly beyond human control, quickly evacuate people to safe areas if time permits to avoid casualties.

8. Self-rescue when getting lost

Firefighters may get lost in a certain place or at a certain time if they can neither reach their destination nor return to the place of departure. It happens sometimes during fighting of forest fires. Such accidents may upset the firefighters mentally, or even cause injury or death. Therefore, care must be taken to avoid getting loss when fighting forest fires.

(1) How to prevent getting loss

① Provide enough education and training and emphasize disciplines during actions: Cadres of grassroots governments, workers, armed police force in forest regions and other fire fighting personnel living and working in forest regions should enhance vocational study, get better knowledge of forest conditions, make a habit of observing relevant principles and take disciplinary actions for those failing to follow regulations and violating principles repeatedly.

② Strengthen management and put in place strict system: Any searching or patrol task in forest should be carried out together by three people, and there should be at least one in them familiar with conditions of the region. During fire fighting and searching on foot, determine the route and the direction beforehand carefully, correct direction when required during actions; when making mistakes, calm down and think carefully before acting again. Check status of radio and intercom and battery condition before leaving and make sure topographic map, compass and GPS are taken to prevent the whole team from getting lost.

③ Enhance field trainings: Organize frequent field trainings to get familiar with and adapt to conditions in forests.

④ Stay vigilant to avoid getting lost: Firefighters living and working in forests must stay alert and avoid getting lost.

(2) How to tell the directions after getting lost (without GPS or compass)

① By observing trees: Isolated trees are better reference. Since trees have a tendency to grow towards the sun, the exuberant side is usually the south. Besides, the sun-facing side has more smooth bark, while the north side is the opposite.

② By observing growth rings: Check stumps in the logging area, or use a saw or a hammer to cut down an isolated tree of over 20 cm when necessary. The side with wider rings is the south and the side with narrow rings is the north.

③ By observing mountains: Sun-facing side of mountains is usually steep with less trees and more barren soil; while the northern side is opposite and grown with moss.

④ By observing rivers: Usually, rivers run from west or northwest to east or southeast.

⑤ By observing the Polaris.

(3) How to keep away from dangers after getting lost

① If a whole team gets lost, avoid reckless actions or decision making all by one person. Team members should stay together and work together and dispersion should be prevented. The team leader should guide members to remember the way here and seek opinions before making decision.

② After determining the way here or the way back, stick to it and ask one member to keep note of mountain shape, terrains and rivers.

③ If you get lost on your own, climb to the top of the mountain as soon as possible and tell the directions by observing where the sun is rising or falling. If night falls, find a shelter from wind and stay away from haunts of wild animals to spend the night. Mentally prepared for the possibility that you may not be able to return for a long time after getting lost. If you find some food, bring some with you after feeding yourself.

④ Watch for lights and sounds at night. Look for people or airplanes during walking. When finding any airplane, find an open ground and wave colorful clothes or lay in on the ground to identify yourself.

⑤ When you get lost, keep to ridges in daytime to observe high facilities, such as watch towers, high-voltage power lines, railways, roads, bridges; light a fire on top of the mountain or at the beach of rivers at night, but take care to keep the fire under control.

⑥ If you get lost on a horse, look after the horse. When marching in daytime, loose the bridle rein to let the horse find its own way. An experience horse is very useful in helping find the way back.

(4) How to organize searching of lost persons: When learning about that some persons are getting lost, fire commanders at all levels should take measures immediately to search for lost persons rather than entertaining the possibility that they will find their way back.

① Well-organized searching: The team dispatched for searching lost persons has to be well-organized. Break up the entire team into groups of at least three members based on local terrains and designate a place of meeting.

② Survey and analysis: Analyze physical condition and mental strength of lost persons and survey the terrain they may be in to determine their scope of activities and possible heading direction.

③ Searching: Search from inside to outside, from small area to large area and from near to far, keep detailed records of observed and searched suspicious circumstances and give timely report to the fire commanders or the command.

④ Indication of direction: There are three ways to indicate direction for lost persons: When the plane finds lost persons, a sketch of routes and directions can be dropped. If dropping is not possible, make left or right banking above the lost persons, and fly in a straight line along the designated direction, then the lost persons may know the right direction. The second method is to light a fire up the mountain. The heavier the smoke is, the better it is. The higher the smoke column is, the better it is, Lost persons can easily see it. The third method is to fire shots at regular intervals. When firing shots, place the muzzle upward to prevent echoes that may mislead the lost persons.

CHAPTER V

Forest Fire Investigation and Archives Management

I. Forest fire investigation

Forest fire investigation includes fire cause, person(s) responsible for fire, burned area, lost of trees and other economic losses. Forest fire investigation not only provides direct evidence for solving fire cases, but also offers firsthand information for selection of revegetation plan in burned areas. Meanwhile, it can serve as materials for summary of forest fire prevention and fighting experience and lessons.

1. Fire cause investigation

After forest fire, its cause has to be identified immediately. There can be many causes of a forest fire, such as continuous dry weather, omissions in management of fire sources, buildup of combustible materials, weak sense of responsibility of fire control personnel. These are all indirect causes. Fire cause as referred in this book refers to the direct cause of forest fire, which is mainly the source of fire. Investigation of fire cause helps to solve the case and call to account of person(s) responsible for the fire. In the long term, it helps to summarize experience, take effective preventive measures and prevent fire reoccurrence. Fire cause investigation is made as follows:

(1) Preparations before investigation: Fire cause investigation should be performed by firefighters or forest police with rich experience. Investigators should arrive at the scene of fire as soon as possible and make necessary preparations, such as bringing notebook, camera, topographic map, marking tapes, compass, measuring instruments and tools, lining rope,

measuring tape, steel rule, tape measure, etc. Get to know transportation and distribution of inhabitants near the scene of fire to enable timely tracing of suspect(s) in case of human related fires.

(2) Interviews: Interview is a key measure to find out fire cause. Objects of interview include persons finding the fire, reporter, persons in the know, related parties, forest rangers and people nearby. Timely records should be made during interview. Interviews should be thorough and careful and time and place of fire, human activities before and after fire, weather conditions should be identified as much as possible to find some clues of fire cause. Meanwhile, such information as fire alarms, mobilization time of the fire fighting team, time of arrival of firefighters and time of fire suppression should be collected in order to identify fire cause and summarize lessons.

(3) Locating fire starting area: Locating the fire starting area (origin of fire) helps to find evidence of fire cause. For large fire scene, shape of fire scars and relevant indications can be used to locate fire starting area to further judge fire spreading direction and identify the origin of fire. While forest fire spreading direction can be told by traces or fire scars on trees, weeds or other indications at the scene. Main methods include:

① Fire coming direction by scars on stumps: Stumps in burned area only burn on one side and scale-shape scars are left. Such burned side can be used to tell the direction the fire came from.

② Fire moving direction by trunk smoking height and tree crown shape after fire: Leeward side of tree trunks usually has higher black burn mark, which means the side with higher smoking height indicates the fire spreading direction. For fire scars left at base of trunks on the slope, if the angel between their upper inclination line and the slope surface is larger than the slope gradient, it means the fire is a surface fire spreading up the mountain along with the wind; if the upper inclination line is almost parallel to the slope, it means the fire is a surface fire spreading down the mountain against the wind. After a forest fire, branches, leaves or bush crowns incline towards the advancing direction of the fire.

③ Fire moving direction by shapes of burned weeds: After a forest fire, vegetation changes its color and decreases in size. Most weeds are turned down and left grass stalks fall to the burned side, which is also the fire coming direction. For clumps of weeds, only the side at the fire coming direction is burned.

④ Fire moving direction by traces left at incombustible matters such as stones: After a forest fire, incombustible matters in burned area are usually more heavily burned on one side, which is the fire coming direction.

(4) Looking for fire cause evidence: After locating fire starting area, it can be divided into several zones for thorough inspection to search for physical evidence of source of fire that may be left, such as cigarette , paper, simple-structured shed, fire pit, candle, match, incense end, incense ash, firecracker, electric wire on the ground, lightning-stricken tree or trunk. If nothing is found, extend the scope of inspection and check repeatedly until evidence is found.

(5) Determining fir cause: If evidence of fire source is found, fire cause can be determined. If fragments or broken pieces of lightning-stricken tree, trunk or root, the source of fire is lightning strike; if cigarette butt, paper or matchstick is found, the source of fire is smoking of cigarette; if simple shed, fire pit, candle, match or food is found, the source of fire is cooking or warming via fire by people in the mountain; if brake shoe, steel piece or charred coal residue from a motor vehicle is found, the source of fire is fire from motor vehicle; if the fire spreads from waste farmland or pond or grassland and it has been established during investigation that someone had been burning grass or the fire control line in such area, the source of fire is grass burning in wasteland or burning of fire control line; if there is tomb, paper ash, incense end or ash, or firecracker, the source of fire is burning from tomb visiting.

Forest fire can be divided into three categories: deliberate, negligent and accidental.

(6) Matters needing attention during fire cause investigation: It is a demanding task to investigate fire cause and find out the person (s) responsible for fire, which has direct effect on handling of the case and dealing with the aftermath. The following matters should be paid special attention during fire cause investigation.

① Fire cause should be investigated in a timely manner. As a rule, after a forest fire occurs, the primary task of firefighters is to put out the fire. At that time, they thought little of protection of fire scene to facilitate case solving. Therefore, the scene of forest fire is usually seriously spoiled. When the commanding officer receives fire report and rushes to the scene, he/she should start considering fire cause investigation. When fire is under control or fire fighting force is sufficient enough, certain police force should be appointed to investigate the fire cause.

② The investigation should be thorough and careful with the scene painstakingly surveyed. For difficult major cases with heavy loss and unclear cause of fire or person (s) responsible for the fire, the investigation should be extended to fully interview residents in nearby villages.

③ Fire cause investigation should go hand in hand with fire case solving. Usually, when a fire breaks out, the person responsible for it will try to put it out; if it is out of control, he/she will run away. Therefore, such person should be controlled before he/she leaves the site as far as possible. If such person has left the scene, determine his/her escape route and direction based on traffic lines and residents near the scene at the same time of fire cause investigation and stop him/her from running away. For fire cases unable to be solved in a timely manner, physical evidence obtained should be handed over to local forest police for further investigation.

④ Physical evidence should be fully used. Evidences of fire cause collected from the scene should be carefully examined and analyzed and relevant theories discussed. The fire scene should be taken photo and recorded in details; a sketch of distribution of the origins and sources of fire should be drawn. Most of the forest fires are man-made accidents, therefore, fire cause investigation is closely connected to fire case solving. Notes from field investigation, physical evidence and testimony of witnesses of fire cause investigators are essential to decision at court.

2. Burned area investigation

Burned area means the sum of different types of area burned during a forest fire, including burned forest, open forest land, shrubland, grassland and wasteland. If a fire occurs in a forest steppe, the fire burns the forest and the steppe at the same time. In this case, the burned area is the total area of burned forest and steppe. In forest fire investigation, burned area is different from affected forest area. Affected forest area refers to the area of burned forest. Forest disaster area refers to the area with dead tree percentage of over 30% in affected forest area or stand area of dead tree percentage of over 60% in young forest. Investigation and measurement of burned area, affected forest area and forest disaster area is an important basis for evaluation of forest fire loss and influence as well as fire case handling.

(1) Surveying methods of burned area

① Visual estimation & plotting: If the accuracy of burned area investigation required is not high, visual estimation & plotting can be used. Such method is performed by experienced investigators by walking around the burned area, plotting its sketch, making estimation of the burned area based on the sketch or estimating directly. This method is applicable when the area of forest fire is not large. If the burned area is large, investigators should walk around the entire burned area, mark major surface features at the boundary of 1:5,000 or 1:10,000 on topographic map and plot the sketch. In mountainous area, investigators can plot at the opposite mountain slope. There is a requirement for boundary displacement error: no more than 0.1 mm for places with visible surface features and geomorphic marks and no more than 0.2 mm for places without visible surface features and geomorphic marks. Error in area should not exceed 7%.

Visual estimation & plotting is not applicable when the burned area is large.

② Field measurement method: If accuracy requirement is high, such instruments as compass, theodolite or total station should be used to measure burned area on site. Traverse and plotting methods are used to plot the plan of the burned area. The following accuracy requirements should be met: traverse error of closure $\leqslant$ 1/200, distance between each measuring station $\leqslant$ 200 m, measuring error $\leqslant$ 1/50, angular deviation $\leqslant$ 1°, area measuring error $\leqslant$ 5%.

③ Aerial mapping method: For large-scale forest fires, an airplane can be used to mark major surface features (rivers, roads, commanding point and buildings) around the burned area on the map and they will be connected to draw the map of the burned area.

④ Satellite mapping method: If the burned area is huge, satellite images can be used to draw the map of the burned area. In general, the ground receiving unit should first process satellite data before calculating different kinds of burned areas. Satellite mapping is accurate and fast.

⑤ GPS measurement: For the burned area of 100–1,000 ha, GPS positioning function can be used to calculate the area in a quick and easy manner. When using this method, investigators take GPS with them, walk around the boundary of the burned area and perform "positioning" operation at every turning point. In this way, the GPS display will provide and save geographical coordinates of each turning point, and then called and used to plot a map in order to define the position and area of the burned area. Some GPS can be used to obtain the area directly. It should be noted when measuring the burned area with GPS that when the terrain around each measuring point is open and unblocked, the measuring accuracy is high; when the measuring points are under tree crowns or inside canyons, the measuring accuracy will be greatly affected.

(2) Calculation of burned area: The burned area can be calculated according to plotted plan or sketch of the burned area. In practice, different methods will be adopted based on shapes and measuring accuracy requirements of burned areas.

① Graph paper method: Place a transparent graph paper or graph paper template on the plan of the burned area and depict its boundary on the graph paper; count the number of intact squares on the paper and add the areas of square fragments to calculate the equivalent number of intact squares, and then calculate the total area based on the map scale. For example, when using a map with a scale of 1/10,000 and a graph paper in cm, the number of intact squares is 22 and the number of square fragments is 20 (the latter is converted into 7 intact squares), a total number of 29 intact squares are obtained; each intact square is equivalent to a real field area of 1 ha, then the total burned area is 29 ha. The procedure should be repeated at least twice with error less than 2%; an average value is obtained after accuracy requirements are met; otherwise, recalculation is required.

② Geometric method: Divide the plan of the burned area into several geometric figures such as triangles or trapezoids, calculate each of them and add up.

③ Planimeter method: Use a planimeter to calculate the burned area in the plan with the pole inside or outside the burned area.

a. Pole outside the burned area: This is a frequently used method in the forestry. When the burned area is not large, the pole can be placed outside it at a fixed point as the origin to record the reading (N_1) from the reading roller (if a planimeter with auto zeroing is used, the origin is "0"), move the tracing point along the boundary of the burned area clockwise for one circle and record the reading again at the origin (N_2). Multiply the difference between two readings (N_2-N_1) with the scale unit (C) of the planimeter to obtain the area P [$P=C*(N_2-N_1)$].

b. Pole inside the burned area: This method is applicable when the area is large and partitioning will be made. The pole must be placed inside the area and the planimeter constant (q) should be added to the reading difference during measuring, which means $P=C*[(N_2-N_1) +q]$ (the values of constants C and q can be found in the attached table of the instrument box).

3. Investigation of loss of trees

(1) Survey Methods: Investigation of loss of trees can be done in two ways: investigation of every tree in the forest and investigation of sample plot.

① Investigation of every tree: If the burned area is small, the forest value is high and cultivation intensity is high, every tree in the forest should be investigated.

② Investigation of sample plot: When the burned area is large and it is difficult to investigate every tree, sample plots can be chosen to estimate loss of trees in the entire burned area. Several typical sample plots should be chosen inside the burned area based on varieties of trees, age of stand, canopy density (degree of closeness), forest type and blasted rate. The sample plots can be set in two ways: strip sample plot with width of 20–30 m and length of 30–50 m in mountainous regions and strip or block sample plot in flat land. The area of the sample plot for investigation should be no less than 1/100 of the burned area and evenly distributed. In sample plots, every tree should be investigated.

(2) Contents of investigation: Investigation of every tree should be made based on variet-

ies and diameter at breast height should be measured for every tree as per destroyed tree, dead tree, burned tree, injured tree and unburned tree. For forest trees with diameter at breast height over 5 cm, their diameter and height should be measured and statistics of their number made to calculate growing stock; for trees with DBH less than 5 cm, only the number is required.

In the practice of forest fire control, blasted rate of trees can be determined according to damages to crown, stem cambium and roots. Burned trees are generally classified into five types to make statistics of the number and volume of timber in order to estimate the total loss of forest trees.

① Destroyed tree: The crown is fully burnt or the stem is seriously burned, rending the tree unfit for use as timber.

② Dead tree: Over 2/3 of the crown is charred or over 2/3 of the stem cambium is damaged (in dark brown) ; the roots are seriously burned and the tree can no longer grow, but it can still be used as timber.

③ Burned tree: Over 1/4 and less than 2/3 of the crown is burned or over half of the stem cambium is left undestroyed; the roots are not seriously burned and there is still room for growing again.

④ Injured tree: Less than 1/4 of the crown is charred; the stem cambium is basically not hurt and only the external bark is blackened by smoke; the roots are not damaged.

⑤ Unburned tree: Tree not burned by the fire at all.

Loss of trees in the entire burned area = burned area * sample plot growing stock/sample plot area.

Some burned trees can still be used. Therefore, certain utilization rates can be used to calculate the percentage of burned trees that can be used as timber.

4. Investigation of burned area types

Investigation of forest fire burned area is closely related to cleanup of the burned area and revegetation. During investigation of forest fire burned area, the burned area can be classified into different types to facilitate revegetation in scientific manner. The burned area can be approximately divided into 5 types based on the degree of damages.

(1) Area with no damage to forest trees: No dead trees are found in such area. It happens when the fire occurs inside forest stands of certain site conditions in particular seasons, time or weather conditions. For example, forest fire cannot reach or spread into picea asperata forest in valleys or dense birch forest.

(2) Slightly damaged area: Only a small layer of barks and soil is burnt inside the burned area; few crowns are burnt; the roots are not hurt; only a small amount of trees with small diameter class are dead.

(3) Moderately damaged area: Tree barks and soil in the burned area are burnt deep; crowns are destroyed; a few roots are burnt; about half of the trees is burnt.

(4) Seriously damaged area: Tree barks are charred; the soil is burnt deep; a certain amount of roots are injured; small branches at the crown are charred; most trees are dead. Such area is usually located in forest stand with artificial young forests or tree varieties with weak fire resistance. For example, in the forest fire on May 6th, 1987 in Greater Khingan Mountains, the death rate of the mature forest is 47.5%, while the average death rate of the young forest is 76.2%. In burned forest stands, only 4.8% of the young forest are left untouched; 51.5% of the young forest are completely destroyed and 43.7% are not completely burnt. As a rule, the death rate at slopes with large gradient is higher than that in flat land. Besides, fires spreading down mountains, crown fires and underground fires are of greater harms. In areas inflicted by such fires, special care should be taken to forest revegetation.

(5) Area with no value: Such area includes waste mountain, grassland or grass pond.
In addition to identification of types of burned areas, observation and study in fixed locations in the burned areas and surrounding areas are required to monitor variation tendency

of vegetation, animals, microorganisms and environmental factors in order to provide scientific basis for revegetation and restoring of forest ecological environment system.

Ⅱ. Cleanup of burned area and revegetation

After a forest fire, measures should be taken to manage the burned area in order to cut forest fire loss. If revegetation measure is not taken, while firewood chopping, wasteland reclaiming and cultivation continue, natural forest will be gradually reduced to secondary forest or even to wasteland. For example, some deserts in Northwest China are formed after forest fires. Minqin County in Gansu Province was once a populated and prosperous place back in Tang Dynasty. However, after a fire forest, this place became desert day after day. Therefore, revegetation in the burned area is the basis of restoring forest ecological system after fire and deserves much attention. Revegetation measures in the burned area include:

1. Cleanup of burned area

Destroyed and dead trees in slightly, moderately and seriously damaged areas should be chopped down and removed. Injured trees with potential of growth should be dealt with according to their conditions. When the degree of closeness of the burned forest stand is over 0.7, seriously damaged trees should be removed first; if it is 0.4–0.6, the degree of closeness should remain no less than 0.4 after logging. Slightly damaged trees or trees with the potential to grow should be retained as far as possible to facilitate natural regeneration. When removing burned trees, the burned area should be completely cleaned up at the same time to clear debris out of the forests to reduce stand flammability.

2. Regeneration with artificial afforestation

Artificial afforestation is required for large-scale moderately or seriously burned area after cleanup and management of tending of young growth should be strengthened. According to investigations after the forest fire on May 6th, the area requiring forest plantation accounts for 20% of the total burned area. In such area, successful restoration of ecological environment depends to a large extent on effectiveness of artificial afforestation.

3. Regeneration with aerial seeding

This measure is applicable to remote burned areas with poor transportation, large burning area and moderate or serious damages. According to investigations of the forest fire on May 6th, in burned areas with all standing forest stock burnt dead (with a total area of 220,000 ha), 30% plots have a burned area of over 60 ha and the biggest ones up to 1,600 ha; 40% and 30% plots have a burned area of 16–60 ha and less than 16 ha respectively. Such plots are short of seed trees and thus artificial afforestation is required to restore forest vegetation. When natural conditions and seedling cultivation ability are poor, aerial seeding is a shortcut of restoring forest resources and ecological environment.

4. Natural regeneration by hillsides enclosure

This measure is applicable to slightly damaged or undamaged burned areas. In such areas, the fire interference does not exceed the stable elasticity limit of the forest ecological system, thus the self-renewal capacity of the ecological system will quickly restore the original forest ecological environment. If the burned area is small and only moderately damaged with seed trees, such measure can also be used. Applicable varieties include larch, Mongolian pine, Pinus yunnanensis and Chinese pine.

5. Artificial promoted natural regeneration

For large moderately damaged and small seriously damaged and burned area, diseased and dead trees should be manually removed and seedlings cultivated with artificial measures to facilitate natural regeneration. White birch and aspen seedlings naturally regenerated after fire in larch forest should be reserved to cultivate mingled forest.

6. Timely cleanup of shrubs to accelerate regeneration

Understory shrubs in the burned area with strong germination capacity can germinate quickly after fire and restore to original degree of coverage within a short time (2–4 years). Therefore, regeneration in the burned area should begin as soon as possible. If the shrubs recover first, it is bad for quick restoration of forest communities consisting of intolerant tree species.

7. Forest pest management

Since the burned area is full of sundries and vulnerable to pests, timely management is required. The number of burned, weak, dying and dead standing trees grows in the burned area, making it prone to xylophagous insects. Therefore, the burned area should be timely cleaned up; trees not seriously burnt and with seeding capacity should be sprayed with pesticide; diseased or rotten trees or topwood can be used as baits to lure pests in and burn them together or use insecticide to fumigate. Pine cone worms or seed pests can be lured and killed with insecticidal sprays.

Ⅲ. Statistics and filing of forest fire

Fire statistics is an important basic job in forest fire control. Since it tells interactions of many factors due to various causes in specific time and place of fire, it can reflect laws and characteristics of fire, which is essential for improving fire prevention and fighting measures, strengthening fire control supervision and improving fire prevention & fighting ability.

Forest fire archives fall to the scope of scientific and technical archives. As original records of forest fire occurrence, spreading, fighting, suppression, investigation, analysis of fire and fire case solving, they are important source of information about fire and key to forest fire study, essential for construction of a modern forest fire control system. By using laws and experience deriving from such archives, forest fire control operation and fire fighting efficiency can both be improved in order to cut consumption of human, material and financial resources.

1. Basic tasks in fire statistics

Basic tasks in fire statistics include statistical survey, processing and analysis of fire occurrence, fighting and handling, provision of statistical data and statistical supervision.

Statistical survey is the first stage of fire statistics. First, collect source data of fire statistics in accurate, timely, comprehensive and systematic manner; Secondly, make statistical processing as per different categories and items to prepare for statistical analysis at the second stage. Fire statistical data is the result of statistical survey, processing and analysis, providing necessary information for forest fire control decision making and forecast.

2. Basic requirement in fire statistics

The basic requirement in fire statistics is to ensure accuracy, uniformity and timeliness of fire statistical data.

(1) Accuracy: Accuracy of statistical figures is vital in fire statistics and authenticity of such figures is the minimum requirement in the work. Otherwise, every task in statistics is meaningless, leading to failure in forest fire control.

(2) Uniformity: Uniformity means that basic data used in fire statistics should meet uniform national requirements. Without uniform requirements, each area and each unit will act on its own as it wishes, thus no uniform summary of the fire can be produced.

(3) Timeliness: Fire statistics should be done in a timely and rapid manner. In addition to submitting fire statistical data to the competent authority as specified, fire characteristics and trends within specific periods should be analyzed, summarized and reported; major, severe and new kinds of fires should be identified quickly and attention of leaders and managers should be called. In this way, the purposes of statistical supervision, service and education can be achieved.

3. Forest fire archives

According to *Regulations on Scientific and Technical Archives*, filing of scientific and technical archives is an essential part of production, technology and scientific study management. Each relevant unit should strengthen management, establish and improve preparation, collection, sorting and filing system of scientific and technical files to ensure integrity, accuracy, systematicness, safety and effective utilization of such archives. Any scientific and technical files requiring filing should be ensured to have legible handwriting and clear drawing to enable long-term retention. According to *Regulations on Forest Fire Control*, the competent forestry departments of the people's governments at or above the county level shall investigate and evaluate along with related departments the fire cause, responsible person, affected forest area and accumulation, casualties and other economic loss of forest fires after fire and then submit the investigation reports to local people's government; the local people's government shall identify and legally deal with units and people responsible for fires based on the investigation reports. Meanwhile, forest fire archives should be established.

According to two Regulations above, forest file archives filing is one of the daily tasks of forest fire offices at all levels. Relevant responsibility system should be established and implemented to ensure every case of fire is provided with sufficient archives.

Fire data is the prerequisite of filing. Without it, nothing can be done. Data of forest fire archives mainly comes from forest fire forecast, reports of hot spots in forest zones by meteorological satellites, fire reports, forest fire fighting information, fire investigation and analysis, fire case solving, statistical data of forest fire and other files, data, minutes, photos, recording, video disks about forest fire. Such data will be collected and processed by secretarial staff or dispatchers in forest fire control offices. Besides, periods of retention of such data should be defined based on their value and their degree of secrecy should be determined as per relevant regulations.

CHAPTER Ⅵ

Quiz of Forest Fire Control Knowledge

Ⅰ. Laws and regulations

1. What are major laws and regulations concerning forest fire control in China?

Major laws and regulations concerning forest fire control in China include: *Criminal Law of the People's Republic of China, Forest Law of the People's Re-public of China, Regulations on Forest Fire Control.*

2. What are the provisions in *Criminal Law of the People's Republic of China* concerning forest fire control?

Article 114. Whoever sets fires, breaches dikes, causes explosions or uses poisonous or radioactive substances, or sources of infectious disease or other dangerous means to endanger public security, if serious consequences have not yet resulted, shall be sentenced to fixed-term imprisonment of not less than three years and not more than ten years.

Article 115. Whoever sets fires, breaches dikes, causes explosions, spreads poisons or poisonous or radioactive substances, or sources of infectious disease or other dangerous means resulting in serious human injury or death or great loss of public or private property shall be sentenced to fixed-term imprisonment of not less than ten years, life imprisonment or death.

Whoever negligently commits the crime mentioned in the preceding paragraph shall be sentenced to fixed-term imprisonment of no less than three years and not more than seven years; if the violations are minor, the offenders shall be sentenced to fixed-term imprisonment of no more than three years or detention.

3. What are the provisions in *Forest Law of the People's Re-public of China* concerning forest fire control?

Refer to the attachment 1.

4. What are the standards of placing a forest fire on file by the public security organ in China?

According to L.A.F [2001] File No. 156 issued jointly by State Forestry Administration and Ministry of Public Security on April 23rd, 2001:

Case of deliberate arson: Any case of deliberate fire setting leading to forest and woods fire should be placed on file. If the burned forest land area exceeds 2 ha, it is a major case; if the burned forest land area exceeds 10 ha or severe injuries or casualties are the result, it is a serious case.

Case of negligent arson: Any case of negligent arson with burned forest land area over 2 ha or leading to severe injuries or casualties should be placed on file; if the burned forest land area exceeds 10 ha or more than five cases of death or severe injury are inflicted, it is a major case; if the burned forest land area exceeds 50 ha or two cases of death are inflicted, it is a serious case.

5. What are the provisions in *Regulations on Forest Fire Control.*

Refer to the attachment 2.

Ⅱ. Knowledge of forest fire control

1. What are the four categories of forest fire warning?

Forest fire warning consists of blue, yellow, orange and red alerts.

(1) Blue alert: Class 2 fire danger appears in an administrative region where combustible materials in forests are hard to ignite, fire is not easy to spread and fire risk is low. In this case, preparation of fire prevention is required.

(2) Yellow alert: Class 3 fire danger appears in an administrative region where combustible materials in forests are relatively easy to ignite, fire is easy to spread and fire risk is medium. In this case, fire prevention starts.

(3) Orange alert: Class 4 fire danger appears in an administrative region where combustible materials in forests are easy to ignite, fire can spread quickly and fiercely and fire risk is high. In this case, emergency fire prevention starts.

(4) Red alert: Class 5 fire danger appears in an administrative region where combustible materials in forests are very dry and inflammable, the humidity is low, temperatureis high and wind is strong, fire can spread very easily and fire risk is extremely high. In this case, urgent fire prevention starts.

2. What does forest fire emergency plan consist of?

(1) Emergency headquarters for forest fire and its functions and duties.
(2) Early-warning and monitoring of forest fire, and the reporting and handling of the information.
(3) Mechanism and measures for emergency response to forest fire.
(4) Safeguards in terms of funds, materials, technologies, etc.
(5) Post-disaster measures.

3. Who should be responsible for costs related to firefighters?

According to Article 45 of *Regulations on Forest Fire Control*: The living allowance and subsidy for loss of working hours, as well as other expenses during the fire fighting period shall, in accordance with the regulations of the province, autonomous region, or municipality directly under the central government, be borne by the unit or individual who caused the fire; if the cause of fire is unknown, the expenses shall be borne by the unit where the fire took place; the local people's government shall pay the portion of expenses mentioned above when the unit or individual who caused the fire, or the unit where the fire took place, is indeed unable to pay. The living allowance and subsidy for loss of working hours, as well as other cost incurred during forest fire fighting can be paid on account by local government.

4. What are common personal injury accidents in fire fighting?

(1) Burns and death by fire. They usually happen when firefighters are trapped in danger and have no time to withdraw. When fire temperature rises to 800–1,000℃, human beings can only survive for 7.5–18 s.

(2) Casualties by suffocation. Carbon monoxide poisoning in fire scene can lead to coma or death. When carbon monoxide content in the air is over 1%, weak victims die in one minute and strong victims can only survive two minutes. When the fire is in the front, victims inhale high-temperature gas and their throats swell, leading to trachea blocking and eventually, death.

(3) Injuries and death by falling. During fire fighting, crashes by broken tree branches, falling stones or falls off cliffs can result in injuries or death.

(4) Casualties by violation of operation procedure. During fire fighting, collision caused by insufficient distance from target objects when using such tools as shovels or rakes, or improper use of fire-extinguishing bullets can result in personal injuries.

5. Who are not allowed in fire fighting?

According to Article 35 of *Regulations on Forest Fire Control*, professional fire fighting teams shall be the major force against forest fire; if the masses are organized to put out forest fires, handicapped personnel, pregnant women, juveniles and other people unsuitable for fire fighting shall not be mobilized to fight a forest fire.

6. What minor features are dangerous zones for fire fighting?

(1) Valleys. When firefighters are suppressing fire in gullies, flying sparks generated will ignite nearby areas and surround the firefighters in fire. When fire burning consumes large amount of oxygen, reduced oxygen content in gullies can cause firefighters suffocate, or even die.

(2) Canyons. When the wind blows along the length of the canyon and the canyon has varied length and width in different positions, wind speed at narrow places increases, resulting in the so-called "canyon effect". When fire burns in the narrow places, fire spreads fast and it is dangerous to fight in canyons.

(3) Branch valleys. When fire occurs in valleys and such valleys have other branches, fire may spread to branch valleys. If fire burns in branch valleys first, it would not easily spread to the main valleys. Therefore, in case of fire in main valleys, it is dangerous for firefighters to advance from branch valleys to main valleys.

(4) Saddles. When the wind blows past ridge saddles (two closely located ridges with small height difference between valleys and ridges), horizontal and vertical cyclones form easily and firefighters can be easily injured.

(5) Ascending mountains. If ascending mountains lay ahead of the fire, the fire can advance quickly and several hilltops can all be burnt within a short time. So it is not safe to form attack line at ridges in front of the fire.

7. What are the techniques used in pneumatic extinguishers?

Pneumatic extinguishers are mainly used to suppress low naked fire, clear fire line and setting barrier zone when fighting fire with fire. Basic techniques used can be summarized as "cut", "press", "top blow", "stir", "sweep" and "dissipate".

(1) "Cut"– cut through flame bottom to isolate burning materials with the flame and sup-

press part of naked fire; at the same time blow small combustion material not fully burnt into the burned area.

(2) “Press”– when flame height exceeds 1 m, use two or more extinguishers to cooperate: raise one up to press flame top so that to lower the flame and push the fire head towards the burned area, creating conditions for extinguisher responsible for fire line cutting.

(3) “Top blow”– when the flame is more than 1.5 m and less than 2.5 m, use several extinguishers to cooperate: raise one to press flame top as mentioned above, use another to blow the middle of the flame to help the first one to lower the flame and push the fire head towards the burned area, while use a third one to “cut” the flame.

(4) “Stir”– in sections in fire scene with heavy forest floor, when the assistances stir the forest floor with a long hook or a long stick with fork, the extinguisher operator push the extinguisher forwards in a trace of lower arc and blow loose small combustion materials into the burned area.

(5) “Sweep”– when clearing the scene, use the extinguisher like a broom to sweep substances not fully burnt in the burnt area to prevent reignition.

(6) “Dissipate”– when suing four or five extinguishers to cooperate, firefighters cannot constantly attack the fire due to high temperature, therefore, one extinguisher can be used to blow the operators at the upper body or head to dissipate heat and reduce temperature in order to improve working environment.

8. What are favorable times of fire fighting?

Favorable times of fire fighting are times when the fire can be easily suppressed. If such times are missed, small fires may grow to large fires. Such times mainly include:

(1) Starting of fire: When the fire just starts, it is weak and small, if the fire fighting team can arrive at such time, it is easy to suppress fire.

(2) Fire spreading down mountains: Such fire is weak and spreads slowly, so it is easy to

suppress. Therefore, try to put it out when it is spreading down mountains as far as possible.

(3) Fire at night: The air temperature is low at night, especially in the wee hours. At such time, relative humidity is high, wind is down, fire is weak, spreads slowly and even extinguishes at lower places. Moreover, fire spreading up mountains advances slowly and even extinguishes by itself. The fire line will break at these hours. In this case, fire can be quickly put out with good commanding. However, since it is dark at night and hard to see the roads, precautions should be taken to prevent falls during fire fighting at night. At the same time, the scene should be carefully cleared to prevent the rise of temperature and increased wind force in the next day which will result in reignition.

(4) Fire in favorable weather conditions: Microclimate in the forest is changeable, so such weather conditions such as cloudiness, rain and snow should be used to put out fire with best efforts.

9. What is "uphill fire"? What is "downhill fire"?

Uphill fire refers to fire spreading quickly from bottom to top of mountains. Downhill fire refers to fire spreading slowly from top to bottom of mountains.

10. What are self-rescue methods in fire?

(1) Withdraw to safe area. When fighting the fire, the fire fighting team should observe changes in fire behavior and, in detecting any jump fire or cyclones, quickly organize firefighter to withdraw to burned areas or areas with less vegetation or lower flame.

(2) Burn down combustible materials in advance as per instructions. Under uniform command, choose flat places, light up head fire while put out fire at both sides and advance along in the direction of fire head to hide in the burned area made by yourselves.

(3) Lie down as per instructions. In case of dangers, lie down at nearby places with less vegetation with feet facing fire coming direction, remove top soil layer on the ground until reaching wet soil, place the face into the soil pit, cover the head with clothes and put hands

in front of the body.

(4) Break through against the wind as per instructions. When the direction of wind and fire suddenly changes, the commanding officer should give orders to break through. Firefighters should act decisively by selecting places with less vegetation, covering the head with clothes, holding breath and running against the wind to break through the fire line. Remember never try to outrun a head fire, instead, run against the fire to break through.

11. How to put out uphill fire?

When fighting an uphill fire, move from the sides along fire heading direction, divide the team into two groups to fight on both sides, advance step by step until reaching the fire head; never try to put out fire head from its front directly.

12. How to put out downhill fire?

Downhill fire spreads slowly with smaller danger and can be easily put out; try best to put it out as soon as possible. In case of downhill fire, the fire head has to be put out first from both sides. If a fire occurs in enclosed hillsides with many combustible materials, gradually put out the fire from sides instead of fighting from a higher position.

13. How to fight when the fire line is in a straight line?

When the fire line is in a straight line, the attack line is long. To put out the fire in shorter time, the team should be divided into several groups to cut the fire line into two or several sections and fight from both ends of each section.

14. How to fight fire when the fire line is in arc or other shape?

“Gnawing-away” method. Fight and advance step by step from several positions. Then break the fire line into several sections, attack from both ends of each section to prevent it from spreading; shorten the fire line until extinguish it in a steady and gradual manner.

15. What should firefighters bring to the scene of fire?

(1) Common fire fighting tools: Pneumatic extinguisher, fire gun, fire mop, fire extinguishing bullet, cleaning tools, etc.

(2) Special fire fighting tools: Mainly used for fire fighting and self-rescue. Each firefighter should bring 2–3 igniters as well as lighters or matches (stored in a damp proof manner) for use in emergencies.

(3) Protective equipment: Flame-retardant clothing, smoke hood, directional light, camping equipment, cooking utensils, foods, medicines, etc.

(4) Other items: Fuels for equipment and tools, quick-wear parts, etc.

(5) Communication device: Brought together by the fire fighting team.

16. How to clear fire scene?

The fire scene should be cleared as follows: Divide the scene into several sections under charge by different people with definite boundaries between them; define tasks and responsibilities for different people; appoint one person to clear about 5–10 m along the boundary from outside to inside to totally extinguish the fire. If remaining fire is found in standing trees, especially those at the boundary of the scene, immediately chop down the trees to put out fire to prevent spreading of slash fire.

ATTACHMENT 1

Excerpts of the *Forest Law of the People's Republic of China* on Forest Fire Control

(Passed in 1984 and amended in 1998)

Chapter I General Provisions

Article 3 Forest resources belong to the state, excluding those specified under law belonging to collective ownership.

Article 4 Forests are classified into the following five categories:

(I) Shelter forests: forests, woods and clusters of bushes with protection as the main aim including water source conservation forests, water and soil conservation forests, shelter forests against wind and for fixing sand, farmland and cattle farm shelter forests, embankment protection forests and highway/railway protection forests;

(II) Timber forests: forests and woods with timber production as the main aim including bamboo groves with production of bamboo materials as the main aim;

(III) Economic forests: woods with the production of fruits, edible oils, drinks, flavorings, industrial raw materials and medicinal materials as the main aim;

(Ⅳ) Fuel forests: woods with the production of fuel as the main aim; and

(Ⅴ) Special-purpose forests: forests and woods with national defense, environmental protection and scientific experiments as the main aim including national defense forests, experimental forests, mother tree forests, environmental protection forests, ornamental forests, woods at ancient and historical sites and revolutionary memorial places and forests in nature reserves.

Article 5 Forestry construction pursues the policy of universal forest protection, afforestation in a big way, combination of felling and cultivation and sustainable exploitation with afforestation as the basis.

Article 11 Tree planting, afforestation and forest protection are the obligation that citizens shall fulfill. People's governments at all levels shall organize voluntary tree planting and afforestation by all citizens and carry out activities of tree planting and afforestation.

Article 12 Units or individuals that have scored remarkable achievements in tree planting and afforestation, forest protection, forest administration and forestry scientific research shall be rewarded by people's governments at all levels.

Chapter Ⅱ Forest Management and Administration

Article 13 The competent departments of forestry at all levels carry out administration and supervision over protection, utilization and renewal of forest resources pursuant to the provisions of this Law.

Article 16 People's governments at all levels shall work out long-term forestry plans. State-owned forestry enterprises, institutions and nature reserves shall compile forest management schemes in accordance with the long-term forest management and submit them to the competent department at higher level for approval and implementation thereupon.

The competent departments of forestry shall guide rural collective economic organizations and state-owned farms, cattle farms and industrial and mining enterprises in the compilation of forest management schemes.

Article 19 Local people's governments at all levels shall organize the departments concerned in the establishment of forest protection organizations to be responsible for the work of forest protection; step up forest protection in the light of actual requirements in the large-area forest regions by building additional forest protection facilities; supervise and urge grass-roots units with forests and those in forest regions to make a forest protection pledge, organize mass forest protection, delimit forest protection responsibility areas and assign full-time or part-time forest guards.

Forest guards may be appointed by people's governments at the county level or at the village level. Main responsibilities of a forest guard are: to patrol and protect forests, and stop acts of destroying forest resources. A forest guard has the power to ask the local department concerned to deal with whomever that has caused destruction of forest resources.

Article 20 Forest public security organs established in forest regions pursuant to relevant state provisions shall be responsible for the maintenance of the social order under jurisdiction and for the protection of forest resources under jurisdiction and may, pursuant to the provisions of this Law and within the authorized scope of the competent department of forestry under the State Council, exercise on its behalf the power of administrative penalties specified in Articles 39, 42, 43 and 44 of this Law.

The Armed Forest Police Force performs the missions of prevention, extinguishment of forest fires and rescue operations assigned by the state.

Article 21 Local people's governments at all levels shall earnestly carry out the work of prevention and extinguishment of forest fires and rescue operations:

(I) Specifying forest fire prevention periods and banning field use of fire in a forest region during forest fire prevention periods; in case of necessity of the use of fire owing to extraordinary circumstances, it must be subjected to the approval of the people's government at the county level or the authorized organ of the people's government at the county level;

(Ⅱ) Installing fire prevention facilities in forest regions;

(Ⅲ) Immediately organizing local army units, civilians and the departments concerned in fire extinguishment and rescue operations in the event of a forest fire;

(Ⅳ) With respect to those injured, disabled or deceased in extinguishment of a forest fire and rescue operations, workers and staff members of the state shall be given medical treatment or pension for the deceased by the units wherein they are employed; non-state workers and staff members shall be given medical treatment or pension for the deceased by the unit where the fire broke out pursuant to the provisions of the competent department concerned under the State Council; where the unit where the fire broke out bears no responsibility for the outbreak of the fire or has no actual ability to bear the burden, the local people's government shall provide the medical treatment and pension for the deceased.

Article 23 Destruction of forest for reclamation and destruction of forest for quarrying, sand gathering and earth gathering as well as other acts of forest destruction are prohibited.

Cutting of firewood and grazing in young forest lands and special-purpose forests are prohibited.

Personnel entering forests and the fringe areas of forests must not shift or damage marks set up in the service of forestry without authorization.

Chapter Ⅵ Legal Liabilities

Article 44 Whoever, in violation of the provisions of this Law, engages in reclamation, quarrying, sand gathering, earth gathering, seed collection, resin collection and other activities resulting in the destruction of forests and woods shall compensate for the losses incurred according to law; the offender shall be ordered by the competent department of forestry to stop the illegal acts, to plant more than 100% and less than three times the number of the trees destroyed, and may be imposed a fine of more than 100% and less than five times the value of the trees destroyed.

Whoever, in violation of the provisions of this Law, cuts firewood or graze cattle in young forest lands or special-purpose forests resulting in the destruction of forests and woods shall compensate for the losses incurred according to law; the offender shall be ordered by the competent department of forestry to stop the illegal acts and plant more than 100% and less than three times the number of the trees destroyed.

If the offender refuses to plant trees or fails to conform to relevant state provisions in planting trees, the competent department of forestry shall do it on his/her behalf, and the expenses required shall be paid by the offender.

Article 46 Functionaries of the competent departments of forestry engaging in forest resources protection and forestry supervision and administration and functionaries concerned of other state organs who abuse power, neglect duty and indulge in self-seeking misconduct constituting a crime shall be investigated of criminal liability according to law; where a crime has not been constituted, administrative sanctions shall be imposed according to law.

Chapter Ⅶ Supplementary Provisions

Article 47 The competent department of forestry under the State Council shall, in pursuance of this Law, formulate measures for its implementation, which should be submitted to the State Council for approval and put into force thereupon.

Article 48 In autonomous regions where provisions of this Law may not be applicable in full, autonomous organs may, pursuant to the principles of this Law and in the light of the characteristics of autonomous regions, formulate flexible or supplementary provisions and submit them to the standing committee of the provincial or autonomous regional people's congress or the Standing Committee of the National People's Congress according to legal procedures for approval before putting them into force.

Article 49 This Law enters into force as of January 1st, 1985.

ATTACHMENT 2

Regulations on Forest Fire Control

(Promulgated by the State Council on January 16th, 1988 and revised and adopted at the 36th Executive Meeting of the State Council on November 19th, 2008 and implemented on January 1st, 2009)

Chapter Ⅰ General Provisions

Article 1 These Regulations are formulated in accordance with the *Forest Law of the People's Republic of China* for the purpose of effectively preventing and suppressing forest fires, ensuring safety of people's lives and property, protecting forest resources and safeguarding ecological security.

Article 2 These Regulations apply to prevention and suppression of forest fires within the territory of the People's Republic of China, with the exception of urban areas.

Article 3 In forest fire prevention, the principle of prevention takes priority and proactive suppression shall be followed.

Article 4 The national forest fire prevention headquarters are responsible for organizing, coordinating and guiding forest fire prevention throughout the economy.

The competent forestry department of the State Council is responsible for supervising and administering forest fire prevention throughout the economy and undertakes the day-to-day work of the national forest fire prevention headquarters.

The other relevant departments of the State Council, in accordance with their respective functions and duties, are responsible for the tasks related to forest fire prevention.

Article 5 The administrative chiefs of local people's governments at various levels assume overall responsibility for forest fire prevention.

The forest fire prevention headquarters set up by local people's governments at or above the county level in light of actual needs are responsible for organizing, coordinating and guiding forest fire prevention within their respective administrative areas.

The competent forestry departments of local people's governments at or above the county level are responsible for supervising and administering forest fire prevention within their respective administrative areas and undertake the day-to-day work of the forest fire prevention headquarters of the said people's governments.

The other relevant departments of local people's governments at or above the county level, in accordance with their respective functions and duties, are responsible for the tasks related to forest fire prevention.

Article 6 Units and individuals managing forests, trees or forest land shall assume the responsibility for forest fire prevention within the areas under their management.

Article 7 Where forest fire prevention involves two or more administrative areas, the local people's governments concerned shall establish a mechanism for the joint prevention of forest fires, demarcate the areas and formulate the rules for joint prevention, in order to share information and strengthen supervision and inspection.

Article 8 The people's governments at or above the county level shall incorporate the construction of forest fire prevention infrastructure into their respective plans for national economic and social development, and include the funds for forest fire prevention in their budgets.

Article 9 The State supports scientific research on forest fire prevention and promotes the application of advanced science and technology to improve the science and technology of forest fire prevention.

Article 10 People's governments at various levels and the departments concerned shall organize regular publicity activities for forest fire prevention, popularize knowledge in this regard and take adequate precautions against forest fires.

Article 11 The State encourages shifting the risk of forest fires through insurance, in order to improve the capacity of the forestry sector to prevent and mitigate disasters and its capacity of self-relief of disasters.

Article 12 Units and individuals that make outstanding achievements in forest fire prevention shall be commended and rewarded in accordance with the relevant provisions of the State.

Units and individuals that perform notable deeds in the suppression of major or severe fires may be commended and rewarded by the forest fire prevention headquarters on the site of forest fires.

Chapter II Prevention of Forest Fires

Article 13 The competent forestry departments of people's governments of provinces, autonomous regions and municipalities directly under the central government shall, in accordance with the criteria for grading forest fire risk zones, as formulated by the competent forestry department of the State Council, grade such zones within their respective administrative areas with the county as the basic unit, publish the results of such grading and submit the same to the competent forestry department of the State Council for the record.

Article 14 The competent forestry department of the State Council shall, on the basis of the grading of forest fire risk zones throughout the economy and in light of the actual needs of work, formulate the national plan for forest fire prevention, submit the same to the State Council or the department authorized by the State Council for approval, and organize the implementation of the plan after it is approved.

The competent forestry departments of local people's governments at or above the county level shall, in accordance with the national plan for forest fire prevention and in light of the local conditions, formulate the plans for forest fire prevention within their respective administrative areas, submit the same to the said people's governments for approval and organize the implementation of the plans after they are approved.

Article 15 The relevant departments of the State Council and the local people's governments at or above the county level shall, in accordance with the plans for forest fire prevention, strengthen the construction of forest fire prevention infrastructure, reserve requisite materials for forest fire prevention and, in light of actual needs, integrate and improve the command information system for forest fire prevention.

The State Council and the people's governments of provinces, autonomous regions and municipalities directly under the central government shall, in light of the actual needs of forest fire prevention, make full use of the satellite remote sensing technologies and the existing military and civil aviation infrastructure to establish a cooperation mechanism for aerial forest protection with the participation of the relevant departments, improve the said infrastructure and secure the funding for aerial forest protection.

Article 16 The competent forestry department of the State Council shall, in accordance with the relevant provisions, formulate an emergency response plan for major and severe forest fires and submit the same to the State Council for approval.

The competent forestry departments of local people's governments at or above the county level shall, in accordance with the relevant provisions, formulate respective emergency response plans for forest fires and submit the same to the people's governments for approval and to the competent forestry departments of the people's governments at a higher level for the record.

People's governments at the county level shall organize the people's governments of the townships (towns) to formulate contingency measures for forest fires in accordance with the forest fire preparedness plans; villagers committees shall, in accordance with the said plans and measures, render assistance in the event of a forest fire.

People's governments at or above the county level and the relevant departments thereof shall organize necessary drills for implementing the forest fire preparedness plans.

Article 17 The emergency response plan for forest fire shall include the following:
(I) The emergency headquarters for forest fire and its functions and duties;
(II) The early-warning and monitoring of forest fire, and the reporting and handling of the information;
(III) The mechanism and measures for emergency response to forest fire;
(IV) Safeguards in terms of funds, materials, and technologies, etc.;
(V) Post-disaster measures.

Article 18 Where an industrial or mining enterprise, a tourist area or a development zone is to be established in a forest area in accordance with law, the forest fire prevention facilities thereof shall be planned, designed, built and inspected for acceptance in a synchronized manner with the said construction project; where tracts of trees are to be planted in a forest area, supporting facilities for forest fire prevention shall be built at the same time.

Article 19 Railway-operating units shall be responsible for forest fire prevention in the forest land under their management, and shall cooperate with the local people's governments at or above the county level in forest fire prevention along the sections of railway lines exposed to the danger of forest fire.

Units responsible for forest fire prevention along electricity lines, telecommunications lines, and oil and gas pipelines shall demarcate forest fire breaks in areas exposed to the danger of forest fire and organize the patrol of the areas.

Article 20 Units and individuals managing forests, trees or forest land shall, in accordance with the provisions of the competent forestry departments, put into practice a responsibility system for forest fire prevention, demarcate responsibility areas, appoint persons in charge of forest fire prevention, and have in place facilities and equipment therefore.

Article 21 Local people's governments at various levels, and State-owned forestry enterprises and institutions shall, in light of actual needs, form professional forest fire fighting teams; local people's governments at or above the county level shall give guidance to units managing forests and to the residents committees, villagers committees, enterprises, and institutions in forest areas in forming community forest fire fighting teams. Both the professional and community forest fire fighting teams shall receive regular training and drills.

Article 22 Units managing forests, trees or forest land shall assign part-time or full-time personnel to take charge of patrolling the wooded areas, controlling the use of fire in outdoor areas, reporting any fire promptly, and assisting the relevant organs to handle cases of forest fires.

Article 23 Local people's governments at or above the county level shall, in accordance with the forest resource distribution and forest fire regularity within the administrative region, delimit forest fire prevention areas, specify forest fire prevention periods and publish the results.

During the forest fire prevention period, the forest fire prevention headquarters of local people's governments at all levels and units and individuals managing forests, trees or forest land shall take corresponding precautionary and preparatory measures based on forest fire danger forecasting.

Article 24 The forest fire prevention headquarters of local people's governments at or above the county level shall organize relevant departments to inspect development of forest fire prevention organizations, implementation of forest fire prevention responsibility system and construction of forest fire prevention facilities by the units concerned in the forest fire prevention area. As for forest fire hazard found in the inspection, the competent forestry department of local people's government at or above the county level shall promptly issue a notice of rectification for forest fire hazard to the relevant units and order them to rectify and eliminate hazards within a specified period.

Units under inspection shall cooperate actively and shall not obstruct or hinder the inspection activities.

Article 25 During the forest fire prevention period, outdoor use of fire shall be prohibited in the forest fire prevention area. If outdoor use of fire is necessary because of prevention on diseases, pests and mice, approval from the people's government at the county level shall be obtained and precaution measures shall also be taken against fire as required; if live-fire drill, blasting and similar activities are necessary in the forest fire prevention area, approval from competent forestry departments of the people's governments of provinces, autonomous regions and municipalities directly under the central government shall be obtained and precaution measures shall also be taken against fire as required; if the Chinese People's Liberation Army and the Chinese People's Armed Police Force shall enter into the forest fire prevention area to deal with contingencies and other urgent tasks, approval from the competent department at the next higher level shall be obtained and precaution measures against fire shall also be taken as required.

Article 26 During the forest fire prevention period, units managing forests, trees and forest land shall set up warning signs for forest fire prevention and carry out publicity activities for forest fire prevention to the people entering their business scope.

During the forest fire prevention period, all motor vehicles entering into the forest fire prevention area shall be installed with fire-preventing and fire-fighting equipment.

Article 27 During the forest fire prevention period, as approved by the people's governments of provinces, autonomous regions and municipalities directly under the central government, the management units of key state-owned forest areas, which are designated by the competent forestry departments and the State Council, may establish forest fire prevention checkpoints to make fire-prevention checks on vehicles and people before letting them enter the forest fire prevention area.

Article 28 During the forest fire prevention period, the local people's governments at or above the county level shall delimit the high-risk forest-fire area and specify the high-risk forest-fire period in case of weather forecasts related to high temperature, drought, strong wind or other high risks of forest fire. If necessary, the local people's governments at or above the county level can issue orders to prohibit outdoor use of fire; and strictly control domestic use of fire that may cause forest fire.

Article 29 During the high-risk forest-fire period, people entering into the high-risk forest-fire area shall carry out activities in strict accordance with approved time, location and scope, subject to the approval from local people's governments at or above the county level and supervision of the competent forestry departments of local people's governments at or above the county level.

Article 30 The meteorological departments and the responsible forestry departments of local people's governments at or above the county level shall set up stations for monitoring and forecasting forest fires, establish the joint consultation mechanism, and timely prepare and issue warnings concerning forest fire risks.

The meteorological departments shall provide weather forecasts related to forest fire risks for free. The radio, TV, press and internet media shall announce or publish in a timely manner weather forecasts related to forest fire risks.

Chapter Ⅲ Forest Fire fighting

Article 31 Local people's governments at or above the county level shall publish the contact number for forest fire alarm and establish the forest fire duty system.

Any unit and individual shall report immediately the forest fire found. After receiving reports, the local people's governments or the forest fire prevention headquarters shall assign staff to check and verify on-site conditions, take corresponding measures to put out fire, and report level by level to the people's governments at the next higher level and the forest fire prevention headquarters subject to relevant regulations.

Article 32 In case of the following forest fires, the forest fire prevention headquarters of the people's governments of provinces, autonomous regions and municipalities directly under the central government shall immediately report to the national forest fire prevention headquarters which shall report to the State Council and its competent departments as specified:

(Ⅰ) A forest fire near a national boundary;
(Ⅱ) A major or severe forest fire;
(Ⅲ) A forest fire that has caused more than 3 deaths or serious injury to more than 10 persons;
(Ⅳ) A forest fire that threatens residential areas or important installations;
(Ⅴ) A forest fire whose visible flames have not been put out within 24 hours;
(Ⅵ) A forest fire in an undeveloped virgin forest;
(Ⅶ) A forest fire threatening the common boundary of a province, autonomous region, and municipality directly under the central government;
(Ⅷ) A forest fire that needs support and assistance from the central authorities to put it out.

The term "more than" as used in paragraph 3 shall include the number itself.

Article 33 In case of forest fire, the forest fire prevention headquarters of local people's governments at or above the county level shall immediately launch the emergency response plan for forest fire as required; in case of major or severe forest fires, the national forest fire prevention headquarters shall launch the emergency response plan for major or severe forest fires.

After launching the emergency response plan for forest fire, the competent forest fire prevention headquarters shall first verify exact location, fire scope, wind force, wind direction and fire behavior, and then reasonably determine fire fighting schemes, divide fire fighting sections, specify and assign person in charge of direct fire fighting operations on site based on weather and geological conditions.

Article 34 The fighting of a forest fire shall proceed under the centralized organization and command of the forest fire prevention headquarters subject to the emergency response plan for forest fire.

Concerned parties shall, under the principles of people oriented and fire fighting with scientific approaches, put out forest fires, evacuate masses threatened by fire and ensure safety of fire fighting personnel to minimize casualties.

Article 35 Professional fire fighting teams shall be the major force against forest fire; if the masses are organized to put out forest fires, handicapped personnel, pregnant women, juveniles and other people unsuitable for fire fighting shall not be mobilized to fight a forest fire.

Article 36 The armed police forest forces shall be responsible for the national forest fire prevention tasks. In case of forest fire fighting, the armed police forest forces shall be subject to the centralized command of the forest fire prevention headquarters of local people's governments at or above the county level; and in case of forest fire fighting across provinces, autonomous regions and municipalities directly under the central government, the armed police forest forces shall be subject to the centralized command of the national fire prevention headquarters.

The Chinese People's Liberation Army shall perform tasks for forest fire fighting in accordance with the Regulation on the *Army's Participation in Disaster Rescue.*

Article 37 In case of forest fires, the departments concerned shall properly perform tasks related to forest fire subject to the emergency response plan for forest fire and the unified command of the forest fire prevention headquarters.

The meteorological departments shall provide weather forecasts and relevant information about the fire area and carry out in good time artificial precipitation operations based on weather conditions.

The competent transportation departments shall give priority to delivery of forest fire fighting personnel and fire fighting materials.

The competent communication departments shall ensure the provision of emergency communication.

The civil affairs departments shall establish shelters and supply centers for relief materials, urgently evacuate and properly settle masses, and carry out disaster relief works.

The public security organs shall maintain security order and strengthen public security management.

The competent departments for commerce and health shall be responsible for materials supply, medical aid, and hygiene and disease control.

Article 38 In accordance with actual needs for forest fire fighting, the forest fire prevention headquarters of local people's governments at or above the county level can decide to take emergency measures, including establishing fire-barrier belts, eliminating obstacles, taking water for emergency use, local traffic control, etc.

Materials, equipment and means of transportation shall be expropriated for forest fire fighting as determined by the people's governments at or above the county level. After extinguishing a forest fire, all expropriated materials, equipment and means of transportation shall be returned in time and appropriate compensations shall also be made according to relevant laws.

Article 39 After extinguishing a forest fire, the fire fighting team shall generally inspect the fire-stricken areas, eliminate remaining fire, and assign sufficient personnel to watch the fire site. Such personnel cannot be withdrawn unless the site is inspected and accepted by the forest fire prevention headquarters of local people's governments.

Chapter Ⅳ Post-disaster Disposal

Article 40 According to affected forest areas and casualties, forest fires are classified into ordinary fires, large fires, major fires and severe fires:

(Ⅰ) Ordinary fires: A forest fire that has affected forest area of less than 1 ha or a forest fire that takes place in other forest lands, or a forest fire that has caused more than 1 death and less than 3 deaths, or serious injury to more than 1 person and less than 10 persons;

(Ⅱ) Large fires: A forest fire that has affected forest area of more than 1 ha and less than 100 ha, or a forest fire that has caused more than 3 deaths and less than 10 deaths, or serious injury to more than 10 persons and less than 50 persons;

(Ⅲ) Major fires: A forest fire that has affected forest area of more than 100 ha and less than 1,000 ha, or a forest fire that has caused more than 10 deaths and less than 30 deaths, or serious injury to more than 50 persons and less than 100 persons;

(Ⅳ) Severe fires: A forest fire that has affected forest area of more than 1,000 ha, or a forest fire that has caused more than 30 deaths, or serious injury to more than 100 persons; The term "more than" herein shall include the number itself and the term "less than" herein shall not include the number itself.

Article 41 The competent forestry departments of the people's governments at or above the county level shall investigate and evaluate along with related departments the fire cause, responsible person, affected forest area and accumulation, casualties and other economic loss of forest fires and then submit the investigation reports to local people's government; the local people's government shall identify and legally deal with units and people responsible for fires based on the investigation reports.

The forest fire loss assessment criteria shall be formulated by the competent forestry departments of the State Council along with related departments.

Article 42 The competent forestry departments of local people's governments at or above the county level shall collect statistics on forest fires as required and submit the same to the statistical organs of the said people's governments and to the competent forestry departments of the people's governments at the next higher level, and also report to related departments of the said people's governments in time.

The forest fire statistics report shall be formulated by the competent forestry departments of the State Council and submitted to the National Bureau of Statistics for records.

Article 43 The forest fire information shall be publicized by the forest fire prevention headquarters of local people's governments at or above the county level or competent forestry departments. Information about major or severe fires shall be publicized by the competent forestry department of the State Council.

Article 44 Personnel who are injured, disabled, or have died from fire fighting shall be treated and compensated for in accordance with the relevant provisions of the State.

Article 45 The living allowance and subsidy for loss of working hours, as well as other expenses during the fire fighting period, shall be in accordance with the regulations of the province, autonomous region, or municipality directly under the central government, be borne by the unit or individual who caused the fire; if the cause of fire is unknown, the expenses shall be borne by the unit where the fire took place; the local people's government shall pay the portion of expenses mentioned above when the unit or individual who caused the fire, or the unit where the fire took place, is indeed unable to pay. The living allowance and subsidy for loss of working hours, as well as other cost incurred during forest fire fighting can be paid on account by local government.

Article 46 After the forest fire, units and individuals managing forests, trees and forest land shall timely take reforestation measures and recover the forest cover of burned areas.

Chapter V Legal Liability

Article 47 In case of any one of the following acts in violation of the Regulations, the local people's governments at or above the county level and their forest fire prevention headquarters, competent forestry departments or other related departments, as well as their personnel shall be obliged to rectify such action under the command of the administrative or supervisory organ at the next higher level; the person directly in charge and other persons directly responsible shall be punished according to laws in severe cases, or be held criminally responsible if the act constitutes a crime:

(I) Failure to prepare the emergency response plan for forest fire in accordance with the relevant provisions;

(II) Failure to issue rectification notice in case of any forest fire hazard;

(III) Approval to outdoor use of fire, live-fire drill or blasting activities inconsistent with the forest fire prevention requirements;

(IV) Concealment, misrepresentation, or willful delay of forest fire reports;

(V) Failure to timely take measures against forest fire;

(VI) Failure to perform legal duties.

Article 48 If any unit or individual managing forests, trees and forest land fails to perform responsibilities of forest fire prevention in violation of the Regulations, such unit or individual shall be obliged to rectify the act under the command of the competent forestry departments of the people's governments at or above the county level; and the individual shall be subject to fines of more than RMB 500 and less than RMB 5,000 or the unit shall be subject to fines of more than RMB 10,000 and less than RMB 50,000.

Article 49 If any unit or individual within the forest fire prevention area refuses to accept forest fire prevention checks or fails to eliminate fire hazards within due time after receipt of rectification notice in violation of the Regulations, such unit or individual shall be obliged to rectify the act under the command of the competent forestry departments of the people's governments at or above the county level; and the individual shall be subject to warning and fines of more than RMB 200 and less than RMB 2,000 or the unit shall be subject to warning and fines of more than RMB 5,000 and less than RMB 10,000.

Article 50 If any unit or individual, within the forest fire prevention period, uses fire outdoors in the forest fire prevention area without authorization in violation of the Regulations, such unit or individual shall be obliged to stop the illegal act under the command of the competent forestry departments of the people's governments at or above the county level; and the individual shall be subject to warning and fines of more than RMB 200 and less than RMB 3,000 or the unit shall be subject to warning and fines of more than RMB 10,000 and less than RMB 50,000.

Article 51 If any unit or individual, within the forest fire prevention period, conduct live-fire drill and blasting activities in the forest fire prevention area without authorization in violation of the Regulations, such unit or individual shall be obliged to stop the illegal act under the command of the competent forestry departments of the people's governments at or above the county level; and the individual or unit shall be subject to warning and fines of more than RMB 50,000 and less than RMB 100,000.

Article 52 In case of any one of the following acts in violation of the Regulations, the unit or individual shall be obliged to rectify the act under the command of the competent forestry departments of the people's governments at or above the county level; and the individual shall be subject to warning and fines of more than RMB 200 and less than RMB 2,000 or the unit shall be subject to warning and fines of more than RMB 2,000 and less than RMB 5,000.

(I) During the forest fire prevention period, any unit managing forests, trees and forest land fails to set up warning signs for forest fire prevention;

(II) During the forest fire prevention period, any motor vehicle entering into the forest fire prevention area fails to be installed with fire-preventing equipment;

(III) During the high-risk forest-fire period, any unit or individual enters into and conducts activities in the high-risk forest-fire area without prior consent.

Article 53 Any unit or individual that has caused a forest fire in violation of the Regulations shall be held criminally responsible if the act constitutes a crime; or be held accountable by law in accordance with Articles 48, 49, 50, 51 and 52 of the Regulations if the act cannot constitute a crime, and the person responsible may be obliged to replant trees under the command of the competent forestry departments of the people's governments at or above the county level.

Chapter VI Supplementary Provisions

Article 54 Special vehicles for forest fire fighting shall, in accordance with provisions, be coated with spraying logo design and installed with alarms and sign lamps.

Article 55 The forest fire occurred in the border area of the People's Republic of China shall be put out in accordance with relevant agreements entered into by and among the People's Republic of China and other countries; or be disposed based on the consultation among the People's Republic of China and other countries if no agreement is made.

Article 56 This provision shall take effect as of January 1st, 2009.

ATTACHMENT 3

National Emergency Response Plan for Forest Fire

(The *National Emergency Response Plan* is revised based on the original *National Emergency Response Plan for Major and Severe Forest Fires* and issued by the General Office of the State Council on December 17th, 2012)

1 General

1.1 Preparation objective

It is prepared to establish and improve a working mechanism against forest fires and make legal emergency response to forest fires in a powerful, orderly and effective manner, thus minimizing forest fires and casualties and property damages, protecting forest resources and maintaining ecological safety.

1.2 Preparation basis

The *Forest Law of the People's Republic of China*, the *Emergency Response Law of the People's Republic of China*, the *Regulations on Forest Fire Control*, the *National Emergency Response Plan for Public Incidents*, etc.

1.3 Application scope

The Plan applies to the emergency response of forest fires occurred in China, except for forest fire in urban areas.

1.4 Operation principle

Emergency responses to forest fires shall be subject to the principles of "unified leadership, cooperation between military and local governments, graded responsibility, territory oriented, people first and scientific fire fighting approach". The administrative chiefs of local people's governments at various levels shall assume overall responsibility. In case of a forest fire, local people's governments at various levels and their related departments shall immediately take measures according to assignment of responsibilities and relevant plans. The people's governments at provincial level shall be the main subjects against major and severe forest fires in their administrative regions, which shall be assisted and supported by the State as required by emergency responses to forest fires.

1.5 Disaster classification

According to affected forest areas and casualties, forest fires are classified into ordinary fires, large fires, major fires and severe fires. Refer to supplementary provisions for standards of disaster classification.

2 Organization and command system

2.1 Forest fire prevention headquarters

The national forest fire prevention headquarters is responsible for organizing, coordinating and guiding forest fire prevention throughout the economy. The general headquarters office is subordinated to the State Forestry Administration and responsible for daily works of the headquarters.

The forest fire prevention headquarters set up by local people's governments at or above the county level in light of actual needs are responsible for organizing, coordinating and guiding forest fire prevention within their respective administrative areas.

2.2 Fire fighting command

Local forest fire prevention headquarters shall be responsible for fire fighting command. Fire fighting against major and severe forest fires across provinces shall be respectively directed by the forest fire prevention headquarters of local people's governments at provincial level, and coordinated and guided by the national forest fire prevention headquarters.

The front-line headquarters shall be established by the forest fire prevention headquarters of local people's governments on the site of forest fire as required. Units and individuals participating in front-line fire fighting shall be subject to the centralized command of the front-line headquarters.

In case of forest fire fighting, the armed police forest forces shall be subject to the command of the forest fire prevention headquarters of local people's governments at or above the county level; and in case of forest fire fighting across provinces, autonomous regions and municipalities directly under the central government, the armed police forest forces shall be subject to the centralized command of the national fire prevention headquarters.

The Chinese People's Liberation Army shall perform tasks for forest fire fighting in accordance with the *Regulation on the Army's Participation in Disaster Rescue.*

2.3 Expert team

The expert team established by the forest fire prevention headquarters is responsible for consultation and suggestion for policies and technologies.

3 Early warning and information reporting

3.1 Early warning

3.1.1 Early warning classification

In accordance with forest fire danger rating, fire behavior features and possible hazards, early warning levels of forest fire can be divided into four grades that are indicated by red, orange, yellow and blue from highest to lowest.

Detailed classification standards for early warning shall be formulated by the State Forestry Administration.

3.1.2 Release of early warning

The competent forestry and meteorological departments at all levels shall strengthen mutual consultation to prepare and publish early warning information for forest fires to the competent departments and the public in affected areas via the early warning information platform, radio, TV program, newspaper, Internet, SMS, etc.

If necessary, the national forest fire prevention headquarters will release early warning information to the forest fire prevention headquarters of local people's governments at provincial level and propose working requirements.

3.1.3 Response to early warning

When the blue and yellow early warning information are released, the people's governments at or above county level and their related departments in warning areas shall pay close attention to weather conditions and changes to early warning of forest fires, strengthen forest fire prevention patrol, satellite forest fire monitoring and lookout monitoring, properly release early warning information and publicize forest fire prevention, improve ignition source management, and properly prepare fire-preventing equipment and materials; and local forest fire fighting teams at various levels shall also be kept on standby.

When the orange and red early warning are released, the people's governments at or above county level and their related departments in warning areas shall, on the basis of emergency response measures for blue and yellow early warnings, further strengthen management of outdoor ignition sources, carry out forest fire prevention inspection, improve broadcast frequency for early warning information, and properly prepare for materials allocation; the armed police forest forces shall also adjust disposition of forces as required and local forest fire fighting teams shall garrison in the front line.

The national forest fire prevention headquarters is responsible for supervising and directing forest fire prevention in the warning areas as required.

3.2 Information reports

The forest fire prevention headquarters of local people's governments at various levels shall report forest fire information in a timely, accurate and specified manner, and timely report to the competent departments in affected areas and the forest fire prevention headquarters in the neighboring administrative regions. In any one of the following cases, the national forest fire prevention headquarters shall report to the State Council immediately along with member units and competent departments of the headquarters:

(1) A major or severe forest fire;
(2) A forest fire that has caused more than 3 deaths or serious injury to more than 10 persons;
(3) A forest fire that threatens residential areas or important installations;
(4) A forest fire that occurs within 5 km from national boundaries or lines of actual control, and also threatens forest resources of China or neighboring countries;
(5) A forest fire threatening the common boundary of a province (district or municipality) ;
(6) A forest fire in an undeveloped virgin forest;
(7) A forest fire whose visible flames have not been put out within 24 hours;
(8) A forest fire that needs support and assistance from the central authorities to put it out;
(9) Other forest fires that shall be reported.

4 Emergency response

4.1 Response at different levels

According to the development trend of forest fires, the fire fighting headquarters and forces shall be timely adjusted under the principles of response at different levels. When a forest fire breaks out, forest fire prevention headquarters at the grass-roots level shall immediately take measures to minimize fire affects as soon as possible. In case of ordinary fires and large fires as initially identified, the forest fire prevention headquarters of the people's governments at the county level shall be responsible for command; in case of major and severe fires as initially identified, the forest fire prevention headquarters of the people's governments at the municipal and provincial levels shall be responsible for command; if necessary, the command levels can be adjusted.

4.2 Response measures

When a forest fire breaks out, local people's governments and competent departments shall organize to take the following measures based on working requirements:

4.2.1 Fire fighting

Grass-roots emergency teams and professional forest fire fighting teams shall be organized immediately to put out the forest fire at its beginning stage. If necessary, local armies, armed polices, militia reserve forces, polices and fire forces shall be called up to put out the forest fire along with fire airplanes and other large-scale equipment. All forces against forest fires shall identify and perform their duties and responsibilities under the unified command of the front-line headquarters. On-site commanders shall carefully analyze the geographical environment and fire conditions, as well as pay close attention to changes of weather and fire behavior during teams advance, station selection and fire fighting operation, so as to ensure safety of fire fighting personnel. Handicapped personnel, pregnant women, juveniles and other people unsuitable for fire fighting shall not be mobilized to fight a forest fire.

4.2.2 Relocation of personnel

If settlements and densely populated areas are threatened by forest fires, parties concerned shall take effective fire retardant measures, formulate emergency evacuation plans, and evacuate residents and affected people in an organized and orderly manner, so as to ensure life safety of masses. In addition, the parties concerned shall also properly relocate masses to ensure necessary provision of food, water, cloths, residence and health care to them.

4.2.3 Treatment of the wounded

The wounded shall be transferred to the hospitals quickly and the seriously wounded may be treated in other places if necessary. The health emergency response teams can be assigned to the fire site as required and the temporary hospitals or medical centers may also be established for on-site treatment.

4.2.4 Post-disaster measures

Parties concerned shall carry out follow-up works for the victims and comfort the bereaved families. Personnel who are injured, disabled, or have died from fire fighting shall be treated and compensated for in accordance with the relevant provisions of the State.

4.2.5 Protection of important targets

If important targets and major hazard sources (such as military installations, nuclear facilities, production and storage equipment of dangerous chemicals, oil and gas pipeline) are threatened by forest fires, parties concerned shall call up professional teams quickly to eliminate threats and ensure safety of targets by arranging isolation belts.

4.2.6 Maintenance of public order

Enhance the social security management of fire affected areas, and crack down on criminal acts such as steal, robbery, plunder of relief supplies, spread of rumors, etc. Strengthen public security patrols around major places (such as financial units and storage warehouses) to maintain social stability.

4.2.7 Information release

Release information about forest fires and response works in a timely, accurate, objective and comprehensive manner by ways of authorized release, news release, press interview, press conference, professional website and official account of Sina Weibo, in order to respond to social concerns. Released contents include starting time, fire-stricken areas, affected area, fire loss, fire fighting procedures, fire case investigation, responsibility investigation, etc.

4.2.8 Site clearing

When extinguishing a forest fire, further organize fire fighting personnel to properly eliminate residual fires, divide responsible areas and assign adequate number of personnel to watch the stricken areas. Fire fighting personnel cannot be evacuated unless there is no fire, fume or gas after inspection and acceptance.

4.2.9 End of emergency

The original department that launched emergency response shall decide to stop the emergency response after extinguishment of forest fires, qualified clearing and acceptance of stricken areas and basic elimination of secondary disasters.

4.3 National response

When a forest fire breaks out, the national responses can be divided into Level Ⅳ, Ⅲ, Ⅱ and I in accordance with fire severity, fire development trend and local fire fighting.

4.3.1 Level Ⅳ response

4.3.1.1 Launch conditions

(1) A forest fire that has caused more than 1 death or serious injury to more than 3 persons, and whose visible flames have not been put out within 24 hours;

(2) A forest fire that occurs in sensitive periods and areas, and whose visible flames have not been put out within 24 hours;

(3) More than three forest fires that occur simultaneously with higher risks.

In case of any one of the above conditions, the national forest fire prevention headquarters will launch Level Ⅳ response.

4.3.1.2 Response measures

(1) The general office of national forest fire prevention headquarters will be on standby to enhance satellite monitoring and timely dispatch fire information.

(2) Strengthen direction on fire fighting operation and mobilize professional forest fire fighting teams from neighboring provinces for supports.

(3) Issue early information for high-risk forest fires based on conditions.

4.3.2 Level Ⅲ response

4.3.2.1 Launch conditions

(1) Major forest fires as initially identified;

(2) A forest fire that occurs in sensitive periods and areas, and whose visible flames have not been put out within 48 hours;

In case of any one of the above conditions, the national forest fire prevention headquarters will launch Level Ⅲ response.

4.3.2.2 Response measures

(1) The general office of national forest fire prevention headquarters shall timely get to

know the latest information of forest fire, organize consultation meetings, research on fire fighting measures, as well as assign working teams as required to assist in coordination and guidance for fire fighting on site.

(2) Forest fire airplanes shall be dispatched from the neighboring units to put out forest fires, in accordance with requirements of the forest fire prevention headquarters of local people's governments at provincial level.

(3) The armed police forest headquarters shall command local armed police forest forces to put out forest fires, and also direct relevant armed police forest forces to prepare for re-inforcement across areas.

(4) The China Meteorological Administration shall provide weather forecasts and real-time weather conditions, and prepare for weather modification operations.

4.3.3 Level Ⅱ response

4.3.3.1 Launch conditions

(1) Severe forest fires as initially identified;

(2) A forest fire that occurs in sensitive periods and areas, and cannot be effectively controlled within 72 hours;

In case of any one of the above conditions, the national forest fire prevention headquarters will launch Level Ⅱ response.

4.3.3.2 Response measures

The following emergency response measures shall be enhanced on the basis of Level Ⅲ response:

(1) The general office of national forest fire prevention headquarters shall organize relevant member units to participate in consultation meetings, research on fire fighting measures and security works, as well as rush to the stricken areas along with personnel from competent departments and expert teams to assist in coordination and guidance for fire fighting.

(2) The armed police forest forces shall be assigned to provide across-area supports, and forest fire airplanes shall also be dispatched from the neighboring provinces (districts or municipalities) to put out forest fires, in accordance with requirements of the forest fire prevention headquarters of local people's governments at provincial level.

(3) Local weather modification operations shall be carried out under direction and supervision based on meteorological conditions of stricken areas.

(4) Allocation and transportation of fire fighting materials and assistance of health emergency teams shall be properly conducted, in accordance with requirements of the forest fire prevention headquarters of local people's governments at provincial level.

(5) Enhance the fire fighting and disaster relief reports along with national media.

4.3.4 Level I response

4.3.4.1 Launch conditions

(1) A severe forest fire that has continuously affected more than 100,000 ha;

(2) Land security and social stability are seriously threatened, relevant industries have suffered heavy economic losses;

(3) The provincial people's government affected by forest fires has no capability and conditions to effectively control the spread of fire.

In case of any one of the above conditions, the national forest fire prevention headquarters will suggest the State Council to launch Level Ⅰ response and the State Council will decide to launch Level Ⅰ response. If necessary, the State Council shall directly decide to launch Level Ⅰ response.

4.3.4.2 Response measures

The national forest fire prevention headquarters shall establish working teams for fire fighting, personnel transfer, emergency support, publicity and reporting, as well as social stability to be responsible for the following emergency measures:

(1) Directing the provincial people's government or forest fire prevention headquarters to formulate the forest fire fighting plans.

(2) Calling up local armies, armed polices, militia reserve forces, polices and professional fire forces from other areas to put out fires along with fire airplanes and other fire fighting equipment and materials.

(3) Arranging domestic supplies, assigning the health emergency teams to treat the wounded and assisting in relocation of affected masses to other provinces (districts and municipalities), in accordance with requirements of the forest fire prevention headquarters of local people's governments at provincial level.

(4) Organizing personnel to repair infrastructures of communication, electric power and transportation so as to ensure emergency communication, electric power supply and smooth traffic for rescue personnel and materials.

(5) Enhancing protection of important objectives and major hazard sources to avoid secondary disasters.

(6) Further strengthening meteorological services and organizing weather modification operations.

(7) Organizing to release forest fire information in a unified manner; collecting analysis and public opinions and directing propaganda, reporting and public opinion for forest fires.

(8) Deciding other major matters related to forest fire fighting.

5 Post-disaster disposal

5.1 Fire evaluation

The competent forestry departments of the people's governments at or above the county level shall investigate and evaluate along with related departments the fire cause, responsible person, affected forest area and accumulation, casualties and other economic loss of forest fires and then submit the evaluation reports to local people's government. The forest fire loss assessment criteria shall be formulated by the State Forestry Administration.

5.2 Work summary

The forest fire prevention headquarters at all levels shall timely sum up and analyze causes of fire and lessons to be learned, so as to propose improvement measures. After extinguishing the severe forest fire, the national forest fire prevention headquarters shall submit the work summary of forest fire fighting to the State Council.

5.3 Rewards and accountability

In accordance with relevant laws and regulations, units and individuals who have made outstanding contributions to fire fighting shall be rewarded, while liable units and individuals shall be held responsible for fire accidents. The personnel who have died from fire fighting and are certified as martyrs shall be treated as per relevant regulations.

6 Comprehensive guarantee

6.1 Team guarantee

Trained and professional forces (such as professional forest fire forces and armed police forest forces) shall be the major subjects against forest fires, supplemented by other forces, militia and reserve forces. If necessary, employees, government officers and masses of the local forestry area can be mobilized to put out forest fires.

Reinforcement across provinces (districts or municipalities) will be carried out as per relevant regulations.

6.2 Transportation guarantee

Reinforcements for fire fighting and their carrying transport equipment shall be mainly transferred by rail transportation or by air transportation via the civil aviation department under special circumstances. Transportation of professional forest fire forces and armed police forest forces shall be ordered by the Ministry of Railways or the civil aviation department as required by the national forest fire prevention headquarters, and carried out by local forest fire prevention headquarters and local armed police forest forces contacting MOR or the civil aviation department.

6.3 Guarantee of fire airplanes

When a forest fire breaks out, first dispatch fire airplanes from neighboring aviation forest protection stations; and then command other fire airplanes within the province (district or municipality) to assist in fire fighting, of which the applications shall be proposed according to the jurisdiction by the provincial forest fire prevention headquarters to northern aviation forest protection station or southern aviation forest protection station under the State Forestry Administration. The protection station will arrange the fire airplanes in a unified manner based on actual conditions.

If the fire airplanes are required to put out major or severe forest fires across provinces (districts or municipalities), the applications shall be proposed according to the jurisdiction by the provincial forest fire prevention headquarters to northern aviation forest protection station or southern aviation forest protection station under the State Forestry Administration. The protection station will arrange the fire airplanes after approval by the State Forestry Administration; if necessary, army planes or other civil planes shall be dispatched as per relevant regulations.

6.4 Communication and information guarantee

Local people's governments at all levels shall establish and improve the emergency communication guarantee system for forest fire prevention, and allocate communication equipment and command vehicles for fire fighting. Communication guarantee departments at all levels shall ensure unblocked communication of forest fire fighting under emergency conditions. The competent forestry and meteorological departments shall timely provide analysis data of weather situation, satellite cloud picture for forest fire monitoring, photo and image of fire-stricken area, electronic map and fire scheduling, in order to ensure auxiliary decision supports for fire fighting.

6.5 Material supplies guarantee

The State Forestry Administration shall, in the key forestry areas, facilitate the establishment of storage warehouses for forest fire prevention materials, and storage of fire fighting tools, protective equipment and communication devices, according to requirements of fire fighting. Establishment of storage warehouses for forest fire prevention materials, and storage of fire fighting tools and equipment shall be conducted by the forest fire prevention headquarters of local people's governments, based on local requirements of forest fire prevention.

6.6 Funds guarantee

The people's governments at or above the county level shall incorporate the construction of forest fire prevention infrastructure into their respective plans for national economic and social development to ensure necessary expenditures for forest fire prevention.

7 Supplementary articles

7.1 Hazard classification criteria

Ordinary fires: A forest fire that has affected forest area of less than 1 ha or a forest fire that takes place in other forest lands, or a forest fire that has caused more than 1 death and less than 3 deaths, or serious injury to more than one person and less than ten persons;

Large fires: A forest fire that has affected forest area of more than 1 ha and less than 100 ha, or a forest fire that has caused more than three deaths and less than ten deaths, or serious injury to more than ten persons and less than 50 persons;

Major fires: A forest fire that has affected forest area of more than 100 ha and less than 1,000 ha, or a forest fire that has caused more than 10 deaths and less than 30 deaths, or serious injury to more than 50 persons and less than 100 persons;

Severe fires: A forest fire that has affected forest area of more than 1,000 ha, or a forest fire that has caused more than 30 deaths, or serious injury to more than 100 persons.

7.2 Overseas forest fires

In case of overseas forest fires (originating from China or neighboring countries), bilateral agreements (if any) shall prevail; relevant measures against forest fires shoud be formulated by the national forest fire prevention headquarters and the Ministry of Foreign Affairs along with corresponding countries if no bilateral agreement has been made.

7.3 Meanings of "more than", "within" and "less than"

The terms "more than" and "within" herein will include the number itself and the term "less than" herein shall not include the number itself.

7.4 Plan management and update

After execution of plans, the State Forestry Administration shall organize plan publicity, training and drills along with competent departments and carry out evaluation and revision to plans in due time based on actual conditions. Local people's governments at all levels shall formulate the emergency response plans for forest fires based on local conditions.

7.5 Plan interpretation

The Plan will be interpreted by the General Office of the State Council.

7.6 Implementation time

The Plan will be implemented as of the issue date.

References

Ba Shuhuan. 2007. Introduction to Forest Fire Prevention and Fighting [M]. Beijing: China Forestry Publishing House.

Bi Zhongzhen, Yao Shuren. 1999. Forest Fire Prevention Works in China [M]. Harbin: Northeast Forestry University Press.

Chen Cunji. 1996. Forest Fire Prevention [M]. Xiamen: Xiamen University Press.

Chen Wengui. 1994. Fire Prevention Management [M]. Beijing: Chinese People's Public Security University Press.

Di Xueying. Wang Hongliang. 1993. Forecasting of Forest Fire [M]. Harbin: Northeast Forestry University Press.

Fan Weicheng. 1992. Introduction to Fire Science [M]. Wuhan: Hubei Science and Technology Press.

Hu Haiqing. 1999. Forest Fire Prevention [M]. Beijing: Economic Science Press.

Hu Haiqing. 2003. Forest Fire and Environment [M]. Harbin: Northeast Forestry University Press.

Hu Haiqing. 2005. Forest Fire Ecology and Management [M]. Beijing: China Forestry Publishing House.

Hu Zhidong. 2002. Forest Fire Prevention [M]. Beijing: China Forestry Publishing House.

Jin Kecan. 1990. Knowledge Quiz of Forest Fire Management [M]. Harbin: Heilongjiang Science and Technology Press.

Kou Wenzheng. 1993. Forest Fire Management System [M]. Beijing: China Forestry Publishing House.

Liu Dejing, Zhai Hongbo, et al. 2009. Construction Standards of Forest Fire Prevention Materials Storage Warehouse (Building Standard 122-2009) [M]. Beijing: China Planning Press.

Liu Dejing, Zhai Hongbo. 2010. Development and Demonstration of Forest Fire Fighting Command and Tactical System [M]. Beijing: China Forestry Press.

Lv Shougao, et al. 1994. Forest Fire Prevention Theory and Technology [M]. Kaifeng: Henan University Press.

Song Zhijie. 1991. Forest Fire Principle and Forest Fire Forecast [M]. Beijing: Meteorological Press.

Wang Dong. 2000. Investigation and Division of Forest Fire Insurance in China [M]. Beijing: China Forestry Publishing House.

Wang Liwei, Yue Jinzhu. 2006. Practical Forest Fire fighting Organization Command and Tactical Technology Reader [M]. Beijing: China Forestry Press.

Wen Dingyuan. 1995. Basic Knowledge of Forest Fire Prevention [M]. Beijing: China Forestry Publishing House.

Xiao Gongwu, Yao Shuren. 1999. Planned Fire [M]. Harbin: Northeast Forestry University Press.

Yang Yulin, et al. 1996. Forest Fire Prevention and Fire Prevention Machinery [M]. Harbin: Northeast Forestry University Press.

Yao Shuren, Wen Dingyuan. 2002. Forest Fire Prevention Management [M]. Beijing: China Forestry Publishing House.

Zhai Hongbo, et al. 2008. Forest Fire Risk Level [M]. Beijing: China Standards Press.

Zhai Hongbo, et al. 2009. Construction Standards of Forest Fire Lookout Monitoring Facilities (Building Standard 123-2009) [M]. Beijing: China Planning Press.

Zhai Hongbo, et al. 2014. Construction Criteria for Comprehensive Management Project of Forest Fire Regions [M]. Beijing: China Forestry Publishing House.

Zhai Hongbo, et al. 2014. The Construction Study on Chinese Key Forest Fire Risk Region [J]. Forest Resource Management (China), 6: 96-103.

Zhai Hongbo, et al. 2018. The Situation Analysis of Chinese Forest Fire Control [J]. China Forest Products Industry, 45(4): 43-48.

Zhai Hongbo. 2015. Practical Manule of Forest Fire Control [M]. Tianjin: Tianjin Science and Technology Press.

Zhang Siyu. 2006. Investigation and Statistics of Forest Fire [M]. Beijing: China Forestry Publishing House.

Zhao Yiting, Zhai Hongbo, et al. 2004. Construction Criteria for Comprehensive Management Project of Key Forest Fire Regions (Trial) [M]. Beijing: China Forestry Publishing House.

Zhen Xuening, Li Xiaochuan. 2010. Forest Fire Prevention Theory and Technology [M]. Beijing: China Forestry Publishing House.

Zheng Huaibing, Zhang Nanqun. 2005. Forest Fire Prevention [M]. Beijing: China Forestry Publishing House.

Zheng Huanneng, et al. 1994. Forest Fire Prevention [M]. Harbin: Northeast Forestry University Press.

Zheng Huanneng, Hu Haiqing, Yao Shuren. 1992. Forest Fire Ecology [M]. Harbin: Northeast Forestry University Press.